Rejecting Climate Doomism

Rejecting Climate Doomism

Diana Stuart

UNIVERSITY OF MICHIGAN PRESS
Ann Arbor

For questions or permissions, please contact um.press.perms@umich.edu

Published in the United States of America by the
University of Michigan Press
First published March 2026

A CIP catalog record for this book is available from the British Library.

Library of Congress Control Number: 2025037144
LC record available at https://lccn.loc.gov/2025037144

ISBN 978-0-472-07797-7 (hardcover : alk. paper)
ISBN 978-0-472-05797-9 (paper : alk. paper)
ISBN 978-0-472-90574-4 (open access ebook)

DOI: https://doi.org/10.3998/mpub.12851428

The University of Michigan Press's open access publishing program is made possible thanks
to additional funding from the University of Michigan Office of the Provost and the
generous support of contributing libraries.

Cover image courtesy iStock / Benjavisa. Design by Shiraz Abdullahi Gallab.

The authorized representative in the EU for product safety and compliance is Easy Access
System Europe, Mustamäe tee 50, 10621 Tallinn, Estonia, gpsr.requests@easproject.com

Contents

Digital materials related to this title can be found on
the Fulcrum platform via the following citable URL:
https://doi.org/10.3998/mpub.12851428

Acknowledgments

I would like to thank the following people for their contributions: Aden Stern for his work editing, Phoenix Eskridge-Aldama for ideas from an earlier paper we wrote that are included in chapter 2, Ryan Gunderson for many fruitful collaborations that helped inform ideas in this book, Brian Petersen for many useful conversations informing discussions in the book, and Brian Emerson for great conversations and suggested readings on changing the way people think and other key concepts in the book.

Introduction

Every Degree Matters

Having taken classes in environmental science in the late 1990s, it seems unfathomable to me that decades later relatively little has been done to address global climate change. Despite targets for reducing greenhouse gas (GHG) emissions, global emissions keep going up. As a result, the average temperature increase over the past decade (what scientists typically refer to) has been about 1.2°C above preindustrial levels. The yearly average temperature for 2024, however, was 1.47°C higher. As we near 1.5°C of warming, we are already seeing unprecedented climate impacts causing significant harm. These impacts will increase in severity with each additional fraction of a degree of warming. Yet it is critical to recognize that failing to meet the political target of staying within 1.5°C of warming does not mean it is too late to act. Much can still be done to avoid our current trajectory of reaching 3°C of warming by the end of the century.

Climate change is increasingly referred to as the climate *crisis* as we see more and more significant impacts. The years 2023 and 2024 were the hottest on record. The summer of 2023 resulted in journalists referring to "global boiling" as coastal waters in Florida and the Mediterranean reached over 30°C (86°F). Related to heat and drought, fires in southern Europe, Canada, and on the Hawaiian island of Maui made global news and resulted in tragic losses. In 2024, new heat records included 117°F in Gaya, India; 120°F in Geraldton, Australia; and 121°F in Aswan, Egypt. Heat waves also lasted longer. Phoenix, Arizona, broke records with 113 consecutive days of temperatures over 100°F. At the same time, major floods decimated areas in multiple countries across Europe. The devastating fires in Los Angeles in early 2025 were attributed to abnormally dry and windy conditions due to climate change. We are already seeing tremendous loss due to global warming, and, unfortunately, we may experience additional warming faster than previously thought. A 2025 study[1] by James Hansen and others suggests that we are not only well on our way to surpassing 1.5°C but we may be closer than we think to 2°C of warming. Yet despite the failures of the past and the significant challenges we face today, action taken now and in the near future can still significantly limit warming.

With current policies in place, we are on track to see temperatures increase 2.7–3°C by 2100.[2] Without these policies in place it would be much worse: The Intergovernmental Panel on Climate Change's (IPCC) latest carbon-intensive scenario reveals the potential to reach from 3.3°C to 5.7°C by 2100.[3] Every additional degree warmer means more heat waves, droughts, fires, floods, storms, hurricanes, famines, migration, and climate-related harm, loss, and suffering. Looking only at temperature-related mortality, one study conservatively estimates that below 2°C the estimated annual excess deaths related to climate change will be around 100,000, but the estimated associated deaths will increase to more than four million at 4°C, and on our current trajectory we would see 83 million excess temperature-related deaths by 2100.[4] This represents an underestimate of total climate-related mortality and the full extent would likely be much more shocking. Yet this level of loss and suffering is not inevitable, as much can still be done to reduce the extent of warming.

Climate change is not all or nothing. While David Wallace-Wells's book *The Uninhabitable Earth* paints a bleak picture of future climate impacts, it also emphasizes that this is not inevitable, that there is still time to avoid the worst climate impacts, and that climate change is far from binary; it is not a matter of being doomed or not.[5] Scientists agree that conditions will get worse with every fraction of a degree of additional warming. While there certainly may be tipping points that escalate impacts, we cannot know when these specific thresholds will be crossed. Staying within 1.5°C is a scientifically supported political target for avoiding increasing levels of climate-related harm. Once this point is crossed, the world will not implode or reach a "point of no return"; impacts will continue to worsen with additional warming. As renowned climate scientist Katherine Hayhoe[6] explains,

> Trying to put a number on exactly how much global temperature change is "dangerous"—and how much carbon we can put into the atmosphere before we hit that level—is like trying to put a number on exactly how many cigarettes we can smoke before developing lung cancer. We know the more we smoke the greater the risk. . . . With both cigarettes and carbon emissions, all science can say is: the sooner we stop, the better.

All evidence points to the fact that every fraction of a degree of warming avoided matters in terms of reducing climate-related harm, loss, and suffering. And, as I will explain in the next chapter, the idea that we are doomed because we are locked in to decades of future warming is not supported by science. In other words, so much can still be done to minimize the extent of global warming. Yet if this is true, then *why have so many people already given up?*

This book will explore this question as well as what we can do to motivate action, what that action must entail, and how it could result in effective climate policies. Many references in this book will be specific to the United States for two reasons: (1) the United States has contributed more cumulative GHG emissions than any other country, making it more responsible for the climate crisis and more obligated to act, and (2) as a US citizen, I feel a tremendous sense of duty to demand government action to minimize warming. While many of the ideas and strategies discussed in this book can be applied in other contexts, the US government and its citizens have a greater moral obligation to act than those in other countries. Yet despite rising concern and support for action, elected representatives continue to resist making the United States the leader it should be in global climate mitigation and adaptation efforts. And while citizens can collectively demand government action, instead we see a concerning rise in climate doomism and inaction.

This book rejects climate doomism and examines how we can create a more livable future and at the same time live more fulfilling lives. There are better alternatives we can work toward, but we need to know that they exist and we must come together en masse to make them happen. Collectively doing nothing simply makes the current climate trajectories a self-fulfilling prophecy. Even with national leaders who do not support climate action, there is much to be done throughout society and at all levels of government. Also, the thinking, planning, and actions taken now will lay the essential foundation for when opportunities for significant policy change arise. There are many good reasons to keep fighting for climate action. Others have argued that it will save significant amounts of money, which it will. Moving beyond the economic reasons, here I focus on a commitment to climate action as a personal choice, a moral obligation, and as an opportunity to live a more satisfying and meaningful life.

1

Rejecting Climate Doomism

Why Climate 'Doomers' Are Replacing Climate 'Deniers'
—*Washington Post* headline, March 24, 2023

Reading about climate doomism and hearing students in my classes express these sentiments, I feel strongly compelled to counter and reject these notions. As the headline above states, climate doomism, or the belief that we are already doomed due to past inaction on climate change, may now be replacing climate denial as a leading excuse for inaction. Yet there is still so much that can be done to reduce global warming and avoid catastrophic impacts. What we need now is more action and more motivation for action. Thus it becomes critical to understand how doomism has become so prominent and what we can do to counter these ideas.

Climate concern is now common, even in places where climate denial campaigns have delayed awareness and concern. A 2023 study by the Yale Program on Climate Change found that 74% of US citizens think global warming is occurring, 66% are concerned about it, and 54% think about it sometimes or often.[1] Yet the same study found that in contrast to levels of concern, much fewer people are talking about climate change or taking personal actions to address it (e.g., 34 and 38% respectively). Sociologist Kari Norgaard[2] coined the term "socially organized denial" to describe the disparity between climate concern and action. Others have described this disparity in terms of "motivational inertia"[3] or "the motivational gap."[4] This lack of agency poses a serious threat to our ability to minimize climate-related impacts and loss.

While I will address other possible reasons for this lack of motivation in the next chapter, here I focus on doomism, and more specifically the defeatist and fatalistic views on climate change that fuel doomism. Doomism represents a state of climate-related despair, where people find no point in acting to mitigate global warming because they have lost all hope that it is even possible to change our current trajectory. Studies show that doomism is widespread, especially among young people. For example, one study[5] on climate beliefs surveyed 10,000 young people (16–25 years old) across 10 different countries and found that 42% reported climate-related despair and 55.7% agreed that "humanity is doomed." Climate scientists and activists alike have recognized that doomism is extremely dangerous as it results in the false belief that it is too late to act on climate change. As I will elaborate on in this chapter, this belief is not supported by science and undermines motivation for action toward the possibilities of reducing global warming and creating a more livable future.

Doomism must be confronted with the truth: There is still so much that can be done to limit warming. It is not too late, and we need to get to work. In the 2023 book aptly titled *Not Too Late*, Rebecca Solnit shares a pointed and timely quote from Mary Annaïse Heglar:

> If you're worried that it's too late to do anything about climate change and we should all just give up, I have great news for you: that day is not coming up in your lifetime. As long as you have breath in your body, you will have work to do.

I fully agree with this view and find it deeply troubling that so many people have succumbed to doomism, especially my students, who have their whole lives ahead of them. Before debunking doomism, I trace some of its origins and identify the perspectives that fuel it.

Doomism: Defeatist and Fatalistic Views

Where did these doomist ideas come from and why do so many people think this way? I first encountered climate doomism when reading Roy Scranton's book *Were Doomed. Now What?*[6] which was published in 2018, the same year as the IPCC *Special Report on 1.5°C*. Scranton represents a prime example of a defeatist who believes that the climate battle is already lost. He argues that "human reason stands defeated" as we remain dependent on an

> extractive fossil-fueled capitalist economy that's killing us, and an elite group of rich and powerful decision makers who believe that they will

be protected from the danger by their wealth, regard flagrant waste and conspicuous consumption as status symbols, and are deeply invested in business as usual even if it means global apocalypse.

In Scranton's view, we cannot successfully challenge these powerful vested interests, therefore we should stop trying and instead accept these forces driving us toward apocalypse. He states that only in our minds and in our thoughts can we escape these forces of impending doom, "which is, in the end, the highest good thought can offer: doing less, doing nothing, being nothing more or less than we are—a gathering of dust and light, a universe—awake." In his view, we are helpless against these forces and must accept our defeat and do nothing, coming to terms with unavoidable climate apocalypse and the insignificance of our own existence.

Accepting defeat and unavoidable apocalypse, however, is very premature. While Scranton draws on philosophy throughout his book, many moral philosophers including contemporaries like Peter Singer and ancients like Aristotle would not agree with the idea that we should do nothing in response to the increasing threats of climate change. Stoic philosophy encourages us to accept the things we cannot change, but climate change is not one of those things. Many of the famous Stoics were deeply engaged in politics and trying to change the world around them for the better. Their lives were far from an example of passive acceptance of the political injustices of their time. They fought against powerful actors to make change. It seems that resignation is attractive to Scranton because it is peaceful and allows him to dwell in his thoughts rather than to act. But with what is at risk, his prematurely defeatist views only undermine the chances of minimizing global warming. If most people agree that the battle is already lost and there is no point in trying, then we will definitely end up losing the battle.

My second encounter with climate doomism was Jonathan Franzen's 2019 article in *The New Yorker* titled "Why Don't We Stop Pretending the Climate Apocalypse Is Coming." Here we see not only a defeatist perspective and the acceptance of apocalypse, but the belief that climate apocalypse is our unavoidable fate. Franzen explains that an "all-out war on climate change made sense only as long as it was winnable." Now it is no longer winnable, he says, because we will surely pass 2°C and "it probably makes no difference how badly we overshoot two degrees; once the point of no return is passed, the world will become self-transforming." Note: This is not true. Instead of focusing on climate mitigation, Franzen argues, since catastrophe is inevitable, we should be focusing more on adaptation. He also states that our time might be better spent focusing on other things we care about. This defeatist and fatalistic position also suggests doing nothing to reduce the extent of global warming.

There was a large negative response to Franzen's article. First, climate scientists stated that Franzen had the science wrong and that there is no "point of no return" at 2°C. Scientists countered that every fraction of a degree past 2°C means more severe impacts, suffering, and loss. Another line of critique focused on Franzen's suggestion that we focus on other things that are meaningful to us. This represents a privileged position as those disproportionately facing social inequities and the full brunt of climate impacts will not have the luxury of focusing on other things when their ability to survive is threatened. Indeed, anyone who thinks we should try not to worry about climate change and "just try to enjoy life" is coming from a position of great privilege, as the global majority will not have this option.

Since this article was published, narratives of climate doomism have increased in the news and across social media. While I do not agree with these defeatist and fatalistic views, I do want to acknowledge the reasons why they might be so appealing. First, I can relate somewhat with the defeatist. This person may have been engaged in climate activism in the past or is a concerned citizen who has voted for candidates who support climate action. Yet after so little has been done over the decades, the defeatist cannot imagine that anything will change and hence the battle has already been lost. Our political leaders remain deaf to the calls of environmentalists and have failed even to end fossil fuel subsidies. There was a battle to be fought years ago, but that battle was lost and now, according to the defeatist, any action is futile. These are understandable thoughts, but I disagree that we have already lost.

We need to redefine our goals. It is too late to avoid climate change. It is already happening, and we must deal with those consequences. But we can shift our goals to focus on *minimizing the extent of future global warming and related harm*. This is possible, and every degree of warming avoided is an achievement. Instead of focusing on what was lost and can no longer be achieved, let's focus on what we can still achieve and the many lives that can be saved and made better through the meaningful mitigation efforts that are still possible.

I can also see how fatalistic beliefs can be strongly aligned with misanthropic views or beliefs that humans have failed to act in accordance with nature and are now paying the price. Many of us who love plants, animals, and being outdoors have likely at one time thought of our own species as a destructive cancer on the Earth, gobbling up resources and hurting other species. Indeed, in some cases our species seems like a bully who takes what it wants when it wants, with no regard for others. These misanthropic thoughts can lead to beliefs that humans are destined for disaster and ruin and that we deserve our fate. As Jonathan Lear describes in *Imagining the End*, after a talk on the trajectory of the Anthropocene, a young scholar stood up and simply stated, "Let me tell you something: We will *not* be missed!"[7] If climate change wipes out humanity, other species may finally be able to thrive or at least recover from the ruin we have caused. Yet the belief that humans deserve extinction overlooks the truth.

Throughout history it has not been *all* humans who have acted like this. In contrast, many Indigenous cultures have long focused on respecting other species and the environment. We see a continued pattern since the colonial era of domination, control, and the forced conversion of people to a competitive, profit-driven system of existence. Many of us now see this system for what it is: one that fails to provide well-being and is driving us toward social and ecological collapse. It is not all humans who have acted like planetary bullies. For the most part, we can blame the relatively few powerful and wealthy humans that have driven and maintained this system, despite its serious harms. And, ironically, if we let humans receive "the fate they deserve," the most severe loss and suffering will be felt by those who have contributed to these problems the least and surely don't deserve this fate. Not all humans have failed to acknowledge ecological realities and acted like planetary tyrants. It has only been some. Unfortunately, they have imposed their views and priorities upon the social and economic systems in our societies.

Lastly, both defeatist and fatalistic views support doing nothing, which may seem appealing as it is easier than doing something. Bringing about the necessary social changes to curtail global warming is going to be difficult. Defeatist and fatalistic views are easier because they require nothing from you; just keep doing what you are doing. Humans tend to resist social and cultural changes, especially when there are many unknowns. Therefore, views that enforce doing nothing, simply continuing with the status quo, are appealing because they are known and because they are easy. As Katherine Hayhoe[8] learned from years of communicating climate science to diverse audiences, it is not the climate science that conservatives disagree with, it is the unknowns about the possible climate solutions.

While denialism was once a leading reason for inaction on climate change, defeatist and fatalistic views may now be more common reasons to justify personal inaction and maintaining the societal status quo. We now have ample evidence that denialism was purposefully propagated to prevent climate action by those vested in fossil fuels and their infrastructure. Are these same vested interests spreading narratives of defeatism and fatalism and disseminating statements of doomism to further prevent action? I wouldn't be surprised. Either way, doomism must be confronted and countered with the truth.

Climate doomism boils down to the notion that doing anything about the climate crisis is futile because it won't make any difference; all roads point to doom. Yet at what point do we reach doom? It might be too late to stay within the 1.5°C target, but a 2°C warmer future will be much safer than a 3°C or 4°C warmer future. Indeed, a 2°C warmer future will be safer than even a 2.1°C warmer future. As every tenth of a degree warmer translates into additional impacts and harm, there is no scientifically backed reason to give up now (see evidence in next section). Still, many people think that if we surpass 1.5°C we should just give up. "ClimateAdam," a climate scientist and YouTuber, compares this situation to a student who is told they cannot receive an A+ on their exam and therefore decides not to study at all even though they could still get an A, B, or C, all better than failing. There are many different climate futures that are still possible, ClimateAdam explains, and doing nothing leads to the worst ones. Staying within 1.5°C of warming was always going to be difficult, but giving up now with so much at stake illustrates a failure to fully grasp the situation.

It is an especially serious concern that climate doomism is becoming increasingly common among young people, as these are the people who have the most to gain from bold climate action taken today. It is also concerning because it can cause serious mental health issues. To live one's life believing we are all doomed and that the world will soon fall apart is not healthy. Some of my students have trouble choosing a career path or deciding if they want to have children because they don't see much of a future ahead of them. Climate anxiety also plagues most young people today and many don't know what to do with these feelings. Defeatist and fatalistic views may offer temporary comfort because they suggest there is nothing that can be done, but in the long term this will do little to quell climate-related fear and anxiety.

Thus, at this moment in history, it is critical to communicate that there is still time to act, that there is still much that can be done, and that each one of us can participate in the necessary efforts to minimize warming. This is what Alaina Wood does on TikTok and Instagram in response to rising doomism in social media. She started telling a different story, the real story that we all need to know and understand, as so much is at stake. More people need to be sharing these messages and countering the dangerous rise of climate doomism spreading through society. Two key communication strategies to counter doomism are: (1) correcting a widespread misunderstanding about the science and (2) highlighting all that can still be saved.

Correcting a Big Misunderstanding

When I asked my class, "What were the main conclusions of the *IPCC Special Report on 1.5°C*?" a student responded, "That if we reach 1.5°C of warming then it is too late and there is nothing we can do to prevent catastrophic warming and social collapse." This is *not* what the report said. The main message was that 1.5°C is a safer target to shoot for than 2°C. Yet it was widely misinterpreted as a "point of no return," where passing this point means it is too late to act. In reality, there is no known "point of no return," and the science behind this has been grossly misunderstood, misinterpreted, and carelessly discussed among prominent figures and in the media. There are tipping points, but crossing one doesn't mean game over.

Scientists have identified up to 15 biophysical tipping points that are at risk of being breached, and it is true that some would be very difficult to reverse, could trigger cascades of impacts, and could result in positive feedback loops of warming. We have not, however, crossed these tipping points, and top scientists studying these possibilities indicate that while tipping points related to the survival of coral reefs are more likely to be breached if we surpass 1.5°C of warming, other tipping points such as the melting of ice sheets are more likely passed at 2°C.[9] The same group of scientists explains that while we are moving toward tipping points with every fraction of a degree of warming, "ambitiously" reducing GHG emission can still limit the risks of crossing these points. Even if one tipping point is crossed (e.g., the death of coral reefs), that does not mean that the world ends. Each tipping point will result in more significant impacts.

All evidence suggests that there is no known "point of no return" and that there is no reason to give up mitigation efforts after temperature increases have surpassed 1.5°C. Scientists don't know if or when tipping points might be reached, but the chances of reaching them will increase with every fraction of a degree of warming. The "point of no return" misunderstanding has become so pervasive that it serves as a primary fuel for climate doomism. Countering doomism therefore requires refuting these misunderstandings.

It is not only students who have gotten the wrong idea. Even famous politicians have stated that if we fail to keep global temperature rise within 1.5°C then we are "damned" or other terms that imply that the world as we know it will come to a sudden end. Journalist Shannon Osaka suggests that the rise of doomism can be traced to the reporting of the 2018 IPCC "*Special Report on 1.5°C*," which focused on staying under 1.5°C and a 12-year target to cut 45% of global emissions.[10] This was widely interpreted as "if we don't stop climate change in 12 years, something catastrophic will happen." This has been refuted and scientists have attempted to clarify this misunderstanding, to little avail. Considering that many projections show a high likelihood of going beyond 1.5°C of warming, it is not surprising that narratives of doomism have spread. As climate scientist Michael Mann[11] argues, this represents a "misunderstanding of science." Staying within 1.5°C was a goal set to prevent significant harm, yet the world will not implode once 1.6°C is reached, and a 1.7°C future is much more livable than a 2.7°C future, which is our current trajectory without much bolder policies to curtail warming. Doomism fails to recognize that global warming is not binary and is far from an all-or-nothing situation.

While doomism has been spreading, far fewer people in the media and among the general public have been talking about positive climate news, such as new climate science that has significant implications for our future. As Michael Mann has widely shared in a variety of media outlets (e.g., *60 Minutes*, *TIME*, *Scientific American*, *The Guardian*), climate science has been revised in recent years based on new models indicating that the lag time between cutting GHG emissions and curtailing temperature increase is much shorter than previous models showed. While models that have emissions staying at today's levels showed continued warming for decades after emissions stop increasing, models looking at reaching net zero emissions show a time lag of only three to five years before warming stops. What this means is that the "lock in" of warming into the future is much shorter than previously thought and that what we do now could result in the stabilizing of temperatures after a few years rather than decades. As a group of climate scientists (including Mann) explained in the *Washington Post*:

> Knowing that 30 more years of rising temperatures are not necessarily locked in can be a game-changer for how people, governments and busi-

nesses respond to the climate crisis. Understanding that we can still save our civilization if we take strong, fast action can banish the despair that paralyzes people and instead motivate them to get involved.[12]

Compared to the idea that we are locked in to decades of warming no matter what, this evidence provides a narrative where people still have a real chance to change the future. In addition, IPCC models show that after net zero emissions is reached, warming will not only stop, but temperatures will slowly go down as natural carbon sinks absorb more carbon from the atmosphere. So gradually, for example, over several decades the average global temperature increase could move from 2°C back down toward 1.5°C.

Not only has the science been misunderstood, which has resulted in the notion that doom is a specific point we will imminently cross, but the science about us not being locked in to decades of warming has been largely ignored. The rise in doomism, therefore, is largely based on a misunderstanding of the science and false assumptions about the realities of the situation. Unfortunately, the hype and fear that we are doomed if we don't act in 12 years (by 2030) spread much more widely across media outlets than any corrections of this interpretation or any revisions in the science. Instead of succumbing to doomism and inaction, the most recent scientific evidence suggests that now is the time to rapidly take action to get GHG emissions down to net zero as fast as possible. When what is most needed is a new sense of urgency and motivation, doomism has done the opposite. Countering doomism thus remains an urgent priority, as the faster we adopt bold climate policies the more can be saved.

Focusing on What Can Be Saved

Another good reason to reject doomism is that there is *so much* that can be saved if we push climate mitigation policies forward, and the bolder the policies the more we can save. First, scientists have used models to predict how many human lives can be saved through different scenarios of climate mitigation. A 2022 estimate from the Lancet Countdown and the Climate Vulnerability Forum[13] shows that we are on a trajectory to experience 3.4 million climate-related deaths *each year* by 2100, but that 91% of mortality can be avoided if warming stays within 1.5°C. Yet even if warming increases beyond 1.5°C many lives can still be saved. As Bressler's 2021 article in *Nature Communications*[14] states, every one million metric tons of emissions reduced in 2020 could save 226 lives by 2100, and cumulatively 74 million lives can be saved if warming can be limited to 2.4°C instead of 4°C by 2100.

We also need to be aware that estimated deaths are not distributed evenly. There will be more temperature-related deaths in Africa and South Asia,[15] where people are the least responsible for climate change and have fewer resources to adapt. The Climate Impact Lab estimates that on our current climate trajectory, by the end of the century mortality rates could reach 106.7 deaths per 100,000 in low-income countries and 25.2 deaths per 100,000 in high-income countries, but that with moderate mitigation efforts mortality can be reduced by 84%.[16]

Lastly, adopting mitigation policies to reduce GHG emissions is estimated to save more than eight million lives simply through reducing air pollution and improving diets (more plant-based) and exercise (walking and biking more instead of driving). We would be better off health-wise with mitigation measures in place. In terms of saving lives, there is every incentive to keep working to reduce GHG emissions and minimize warming, even if warming surpasses 1.5°C.

In addition to human mortality, global warming also increases unemployment, poverty, and other forms of social precarity, especially in the Global South and for those who are already struggling to get by. This means losing the chance to live a life where the basic needs of food, water, and shelter are met. Beyond basic needs, climate change can also curtail possibilities for people to have what is necessary to live a good or flourishing life. There are other things people need to have a decent life. For example, Aristotle describes the human needs for friendship, knowledge, honor, virtue, and pleasure. A more modern adaptation is Martha Nussbaum's[17] central capabilities approach, where all humans should have the ability to meet their needs related to life, health, imagination, emotions, play, practical reason, connections with others, and influence over one's environment. Humans need more than their basic needs met to thrive. Most parents prioritize meeting their children's basic needs but also want them to have opportunities to develop who they are, express themselves, connect with others, and flourish. Climate change threatens not only life but the chances of having a good life. Every degree of warming we can avoid gives more people a chance to live a decent life.

If you care deeply about plant and animal species like I do, mitigation efforts are also still worth aggressively pursuing even if the average global temperature increase surpasses 1.5°C. Every degree of global warming impacts species through further shifting the geographical ranges of habitats and by increasing threats from extreme heat, drought, heavy rains, and flooding. In addition, the more glaciers and ice that melts, the more arctic species will lose their habitat. As conditions change, organisms also become more susceptible to disease and other mortality risks. The IPCC estimates that 9–14% of species would be at very high risk of extinction at 1.5°C of warming, 12–29% at very high risk at 3°C, and 15–48% at very high risk at 5°C.[18] Evidence suggests that warming beyond 5°C would result in mass extinction. Minimizing warming is the best way to preserve biodiversity, but every degree of warming that can be avoided gives species a better chance at survival.

Many people also talk about how much money can be saved if we boldly invest in mitigation efforts now and avoid extensive future climate damage that can be costly in terms of property, commerce, and health. Monetary costs of future climate impacts are a bit complicated because they are not simply straightforward estimates of future costs. They are discounted. In economics, discounting places less value on future people and their preferences. This is done because it is assumed that future generations will be wealthier and better off and that the desires of today's generation should be prioritized. This matters because discounting reduces the costs and benefits calculated for future citizens and prioritizes reducing costs and maximizing benefits for current citizens. Many have critiqued discounting for being inaccurate and immoral, as it results in underestimates of costs and benefits for people in the future.

But even with the economic decks stacked against bold mitigation action now due to discounting, the number crunching reveals tremendous cost savings if governments do adopt these measures. A study focused on the United States estimates that if warming is limited to 2°C the country would avoid $1 trillion in damages by 2050 and $8 trillion in climate damages by 2100, and that every degree of warming translates into losing 1.2% of GDP.[19] Other global estimates range up to $140 trillion in savings from moderate mitigation efforts.[20] Even though estimates show a reduction in GDP early on in mitigation efforts, the cost savings from avoiding climate damage far surpass the initial economic losses. In addition, savings may be even greater because current analyses tend to focus more on the costs of mitigation actions while overlooking many of the monetary benefits of climate action.[21]

As mentioned earlier, Bressler estimates that every one million metric tons of emissions reduced now would save 226 lives by 2100, but he also estimates that every one million metric tons of emissions reduced would reduce climate damage costs by $258 million. Bressler, however, warns against using such estimates in general because they attempt to turn all types of damages into monetary values and future damages are again discounted based on the assumption that the welfare of people in future generations is worth less. Bressler states these are "subjective ethical choices" and because of this "the best practice should be to provide estimates of the non-market effects of emissions in original units [e.g., mortality] without monetization or discounting." Since estimates with discount rates already reveal the net financial gains of climate mitigation, we can surely know that without these "subjective ethical choices" in cost-benefit analysis the evidence would even further show the tremendous economic benefits of adopting bold climate policies as soon as possible.

What remains problematic is that most people are not planning their finances over the next 80 years and are focused on shorter time horizons, and if we look at who loses money through the stranded assets associated with bold climate mitigation it explains why fossil fuel companies and related interests continue to use their political connections to impede mitigation efforts. Therefore, despite the many benefits of action, gridlock remains due to those vested in and benefiting from the current system. This gridlock is not insurmountable, but it must be confronted. There is so much to save and no reason to give up now, but certain interests would rather that we did give up and let them continue to reap the profits from the status quo. Considering what could be saved/lost and for whom, the climate protestors with signs saying "People over Profits" are definitely on to something.

Way Too Early for Doom and Despair

Given that the scientific evidence shows that every bit of additional warming translates into more suffering and loss, and considering there is still so much that can be saved even if we surpass current climate targets, it is way too early for doom and despair. Doomism perniciously undermines the urgently needed motivation to act and must be confronted and countered with the truth. If we get world leaders to adopt bold climate policies, it can save millions of lives, millions of species, and trillions of dollars. Doom and despair result in inaction. With so much at stake, why would anyone give up now?

Doomism, as a state of climate-related despair, is also bad for our mental health and will undermine our personal chances of living a good or flourishing life. As philosopher Brian Treanor[22] explains:

> [D]espair is fatal to both environmental progress and individual flourishing. . . . It is fatal to environmental progress because while it is true that we may not be able to adequately respond to certain crises in time to avoid their negative effects, failing to try ensures failure and often exacerbates the situation. Despair is fatal to flourishing because it undermines our belief in the significance of our actions and our lives.

Rejecting doom and despair increases the possibilities for positive social-ecological outcomes and it also supports personal well-being.

What we need instead of doomism is more motivation to act, even in the face of setbacks and clear challenges. As I will discuss in chapter 3, the "climate action" most necessary is political engagement to pressure governments to create and push forward bold mitigation policies that can minimize global warming. This will not happen overnight, but it is the only way to reduce the majority of emissions driving us into an increasingly warmer future. We need more political engagement and collective action, yet doomism simply undermines all of this. Even in situations where national leaders oppose climate action, much can be accomplished in other levels of government and work done now is critical for laying the necessary groundwork for rapid and meaningful mitigation actions when the political opportunity arises. If we reject doom, we can instead focus on what can motivate more people to participate in climate activism, what kinds of new policies should be demanded, and how working together to address the climate crisis represents an opportunity to increase our personal well-being while creating a more just and livable future.

2

Motivating Climate Action

The helplessness we feel is a sham. Every decision counts, and from every moment forward each choice has the potential to contribute to the prevention of more suffering.

—Britt Wray

When my eight-year-old daughter asked me, "Should I be afraid of climate change?" it was difficult to know what to say. While I don't want her to live in fear, the truth is that the realities of climate change are scary, and there are good reasons to be afraid. Yet fear is a funny thing. For some people it can motivate action to reduce risks and for others it can lead to denial, paralysis, and inaction. I think there are other important motivators beyond fear as well as shame or guilt. There are many positive things to be gained from climate action. A positive focus can provide a strong motivation to act and is much more likely to help sustain the collective action necessary to demand bold climate policies. In this chapter, I first examine additional reasons beyond doomism that help explain why motivation to act has been lacking. I then argue that negative emotional motivators are not the only way to spur collective action, and I discuss how some positive aspects of climate action can offer additional reasons to motivate and sustain action.

A Crisis of Agency

As stated in the previous chapter, there remains a significant disparity between climate concern and climate action. Chapter 1 examined doomism as one explanation of this disparity. But before doomism emerged, other reasons for this motivational gap or motivational inertia had already been identified. Despite being educated, concerned, and anxious about climate change, individuals had other reasons not to act.

One barrier to motivation is a common misperception about public support for climate action. This relates to what is called "pluralistic ignorance" in the psychology literature, which refers to a widespread shared misperception about what the majority of other people think. In the context of climate change, most people assume that the rest of the population is not concerned about climate change and would not support climate policies, like a carbon tax, 100% renewable energy, or a Green New Deal. According to a 2022 study by psychologists,[1] approximately 80–90% of Americans underestimate public concern about climate change and support for far-reaching and transformative policies. The researchers found that a supermajority of US citizens supports bold climate policies (66–80% depending on the policy). But due to pluralistic ignorance, the dominant belief is that only 3% of the population supports climate policies. As the researchers state, "supporters of major climate policies outnumber opponents 2 to 1, but Americans falsely perceive nearly the opposite to be true." In fact, majority support for climate action is global, with a 2023 study[2] finding that across 23 different countries, 77% of people agreed with the statement, "It is essential that our government does whatever it takes to limit the effects of climate change." Only 10% of respondents disagreed with the statement.

False assumptions about climate beliefs are significant because it reduces the likelihood that people will talk about climate change and climate policies, and it undermines motivation to act to demand bold climate action. People are much less likely to talk about their opinions if they think their views are not shared by others, something researchers in a 2023 article[3] call the "spiral of silence." This creates a context in which people refrain from talking about climate change and it remains a topic that is not socially acceptable for conversation. The spiral refers to the cycle of silence that this pattern creates. In addition to not talking about the seriousness of climate change, this pattern curtails constructive conversations about policy and collective action. Feeling like they are alone in their views on climate, many are also discouraged from seeking out collective action. If we falsely assume we are the only ones who care, we are much less likely to act.

In *Generation Dread*, Britt Wray explores the emotions behind climate inaction and discusses the role of negation and disavowal. Negation is the act of pushing reality away and repressing something, like climate change, from our thoughts. This allows us to ignore it and carry on as usual. Disavowal, Wray explains, is a soft kind of denial; it is like "having one eye open and one eye closed at the same time." Most people believe that climate change is real, they believe in the science, and they know the risks, but many downplay the extent of risks and threats so that they can continue with their daily lives. These tactics are used as a defense mechanism, as many people feel helpless to act on climate change. But Wray states, "What's weird about the learned helplessness we see in society regarding this threat is that we've never really failed because we've never really tried. Our narrative of powerlessness is holding us back more than our actual capacities." Believing we are helpless leads to denial, negation, and disavowal, which all demotivate action.

A widely cited study by Markowitz and Shariff[4] identifies six psychological aspects of climate change that challenge the human moral judgment system and fail to motivate action. Many of these apply to individual behavior change as well as political engagement. This includes the spatial and temporal distance of impacts, which are taking place in "long time horizons" and in "faraway places." Others include the unintentionality of climate harm, the abstract and complex nature of climate change, self-defensive biases (it's not my fault), unrealistic optimism in climate solutions (e.g., techno-fixes), and political polarization. As climate impacts hit closer to home, notions that it takes place over long time horizons and in faraway places will no longer represent as much of a challenge. Also, focusing on government-enacted climate policy rather than individual behavior changes could reduce some self-defensive biases. Still, with so many lives at stake, it is puzzling that so many people fail to see climate inaction in moral terms. Markowitz and Shariff conclude that "right and wrong are powerful motivators," but "this source of motivation for compassionate and forward-looking action, including support for both political and behavioral change, may be largely untapped in the context of anthropogenic climate change." In other words, more people should be talking about climate change in moral terms as this could significantly increase climate action.

Writing about the ethics of climate change, philosophers Dale Jamieson[5] and Stephen Gardiner[6] disagree on the fundamental causes of continued public inaction. Jamieson argues that despite the moral responsibility for people in wealthy carbon-intensive countries to act, many people fail to see inaction as immoral because it is difficult to identify a direct perpetrator and victim. Therefore, it does not register as immoral. Gardiner, on the other hand, argues that people do (to some degree) understand that inaction is immoral, yet they find ways to rationalize their position or simply fail to care. In other words, people either fail to see climate change as a moral issue, rationalize away the immoral implications, or do see it as immoral but simply don't care. Despite their differences, what Jamieson and Gardiner agree upon is that we need new ways to motivate an adequate response to the threats of climate change. As stated by Jamieson, "[p]erhaps the greatest challenge of this century is to reconstruct and insatiate forms of individual and collective agency."

Jamieson argues we have a "crisis of agency" in terms of both individuals and our political institutions. The ways we cognitively understand and talk about climate change fail to result in the collective action necessary. Some argue that climate inaction is so pervasive because the issue has thus far failed to engage our emotions in a way that would motivate sustained action. What about this crisis has failed to engage our emotions and, now that the impacts are becoming more obvious and severe, are there new ways of thinking about climate change that can better engage emotions and motivate action?

Beyond Negative Motivators

Potential catastrophic scenarios can be powerful motivators for political action. Given current and future climate change scenarios, it is curious that more people are not taking action. One explanation is that climate change remains a lower priority compared to other issues such as the economy and health care. Continued high-level fear is also difficult for people to sustain. Instead, climate anxiety, a lower level of fear, is more common. In some cases, though, climate anxiety does not lead to action. For some individuals these emotions could result in not wanting to acknowledge something so upsetting and therefore withdrawing from thinking about or acting on the problem. In addition, people may deny and repress climate-related fear and anxiety because they do not feel there is an effective route for change. Many avenues for action seem closed off due to current power dynamics, creating a sense of helplessness.

Another negative emotion that can drive climate action is guilt. This includes guilt related to personal choices or group-based guilt. Group-based guilt is when individuals see the responsibility their group has for a negative outcome and feel guilty because they are part of that group. This could apply to being a citizen of a wealthy country that is most responsible for cumulative GHG emissions. It could also apply to intergenerational group-based guilt, or the notion of having let future generations down. Guilt can also be associated with the idea of bystanding, or failing to act when others are being harmed. Other examples of bystanding include doing nothing in the face of racism, slavery, murder, or genocide, where bystanding more clearly makes one complicit in the harm done. As Dante put it, "The hottest places in hell are reserved for those who in times of great moral crises maintain their neutrality." If inaction on climate makes us complicit in the outcomes, this indeed is a heavy moral burden. Feeling guilty can motivate some people to act to reduce future harm, but it could also result in defensiveness, backlash, or withdrawing from the issue altogether.

Another emotion that can motivate climate action is anger. Anger is often called a negative emotion, yet in some cases expressing anger can also be a beneficial experience. Eco-anger, or anger at the parties causing environmental damage and destruction, is a common sentiment and applies to those angry at governments and fossil fuel companies for failing to address the climate crisis.[7] Anger can be a key motivator for climate activism, even more so than anxiety, and can also help improve mental health outcomes.[8] In this way anger can be a positive emotion as it results in a personal catharsis and drives proactive measures to challenge power. It is likely unhealthy, though, for someone to be angry all the time. For some people, anger might be a key motivator to initiate climate activism, but additional emotional motivators may be necessary to sustain action.

Negative emotions are understandable given the climate trajectory, and in many cases they do motivate action. As Britt Wray[9] explains, "even the smallest degree of eco-distress, when harnessed, also has the potential to spur change for the better." Allowing oneself to experience and move through the negative emotions one is feeling is also important for our well-being. Yet beyond fear, guilt, anxiety, and anger, other more positive emotions can motivate and especially sustain climate action. Cultivating positive emotions and focusing on the rewarding and beneficial reasons to act may result in more widespread and sustained actions.[10]

Next I examine some motivators for climate action that are active, positive, rewarding, and beneficial in ways that might not only increase and sustain participation but could help improve our day-to-day lives. Guilt, fear, anxiety, and anger are understandable but will only get us so far. We also need reasons to act and mindsets that help us to feel good, act justly, and live the most positive lives we can while working together to minimize warming. What might be some good reasons to motivate climate action? How about protecting the future of today's children, fighting for a more just world, living a good and virtuous life, experiencing the joys of working with others on something truly important, and helping to build a better world that is more sustainable and just? Focusing on these reasons not only can help motivate and sustain action, but helps us to live a more meaningful, satisfying, and fulfilling life.

Protecting the Future of Today's Children

Climate change analyses and discourse often emphasize impacts on future generations. For example, a 2023 poll found that 71% of Americans think climate change will harm future generations.[11] But while climate impacts will get worse in the future, we need not focus on abstract future generations when climate change will significantly impact the lives of people alive today, especially children. These children already have names, faces, and personalities. Without significant interventions, these children will face extreme climate impacts in their lifetime. Concern for future generations made more sense a few decades ago. But due to inadequate mitigation efforts, climate change is a current and near-future threat. Focusing attention on young children, rather than on future generations, could help to emotionally motivate climate activism. In other words, while temporal and spatial abstraction may represent a motivational roadblock, one need not look further than the face of any child to see a person whose lifetime will be shaped by climate-related disasters, loss, and suffering.

One can easily draw from a wealth of climate science to understand the likely impacts that today's children will experience. Future climate impacts are typically discussed in terms of projections, which vary based on the trajectory of current mitigation efforts. For example, the Representative Concentration Pathways (RCPs) assess the impacts of climate change based on four developmental pathways associated with different levels of radiative forcing, and the Shared Socioeconomic Pathways (SSPs) represent climate impacts in five alternative scenarios of development. Climate impact analysis and projections from the RCPs and SSPs often provide data extending to the year 2100. Most adults alive today will not experience the climate change impacts expected by the end of the century, but children will. We need not list the range of horrors that these impacts might entail, as you likely already know.

Regardless of whether a young child currently resides in the United States, Japan, Nigeria, Ecuador, or India, the severity of the effects of climate change they experience will be largely determined by socioeconomic factors. Depending on their financial resources, race, gender, education, employment, and abilities, they might be able to avoid or adapt to some of these projected impacts, but not all. In general, children in wealthy countries will have more resources, social protections, and options. Some of these children will face unprecedented challenges that may result in cutting their lives short or undermining their basic needs, yet all will face some loss as climate change curtails their opportunities for a stable and flourishing life.

As climate scientist and activist James Hansen[12] explains, "[c]itizens with a special interest in their loved ones need to become familiar with the science." Focusing on abstract future generations obscures the reality that our loved ones, especially those who are very young, face an extreme and hazardous future. In contrast to focusing on an abstract idea of future generations, focusing on protecting the children of today represents a more tangible notion to emotionally connect to and could increase positive motivations to act. Even if this focus only motivates those who are parents, they alone would represent a potentially powerful group to demand their governments take bold climate action.[13]

Standing Up for Justice

Peter Singer, a contemporary moral philosopher, presents us with a scenario that we can apply to climate inaction. If you are walking by a pond and see a small unattended child drowning in the pond, you can keep walking and let the child die, or you can run into the pond and save the child. The rescue might cost you something. You might ruin your best suit or pair of shoes, or your wallet may fall into the water. Yet, he argues, you are still morally obligated to save the child. Singer even argues that if you have the means to save someone you cannot even see but you know needs help somewhere, you are still morally obligated to act.

Instead of thinking of this in terms of individuals, let us redirect these ideas to nations. Rich countries are most responsible for climate-driven harm and that means in our analogy they are the reason the child faces death. In this case, say you were riding your bicycle on the way to work and didn't see the child playing by the pond and accidentally knocked the child into the water. Because you were the cause of the child falling in, you would most likely feel even more morally obligated to act. Like the commuting cyclist, wealthy nations know they are responsible, have the means to intervene, and know they can save lives. Even if we can't directly see people at risk of death right in front of us, we know they exist and are facing increasing climate-related threats. But acting will cost the rich nations something. What should they do?

Any sense of justice requires that those who have caused harm, in this case emitted the most GHG emissions, are responsible for repairing that harm. Based on the polluter pays (those who polluted are at fault), the beneficiary pays (those who have benefited are at fault), or the ability to pay (those with the most resources pay) principles, political theorist Catriona McKinnon states that "all roads lead to Rome." This means that whatever method of assigning responsibility one choses, rich countries are the most responsible and should be the ones paying for rapid mitigation and adaptation. As low-income countries face the most severe climate impacts, high-income countries are morally obligated to pay for mitigation and adaptation efforts in these countries.

If wealthy countries are at fault, rather than focusing on a sense of obligation or guilt, why not focus on the action of repairing harm and saving lives? Returning to Singer's child in the pond scenario, you can feel guilty and obligated to save the child and bummed out about ruining your clothes, or you could not give it a second thought and simply dive into the pond to rescue the child because helping those who are facing threats and saving lives is the right thing to do.

Reducing climate-related harm in poorer countries is also an act of justice. What is justice? Justice goes beyond utilitarian accounting that would favor actions that result in the greatest overall good or average positive benefits (in terms of preferences, pleasure, etc.) and instead looks at distribution and the fairness of actions. Philosopher John Rawls[14] has famously argued that it is the duty of our political institutions to uphold justice in terms of fairness, focusing on first meeting citizens' basic needs. Further, according to Rawl's difference principle, resource distribution should prioritize improving the circumstances for those who are most disadvantaged. In other words, those who are at the bottom of the economic ladder and who struggle to have a decent life should be at the top of the priority list for the benefits they need.

The negative impacts of climate change are felt most severely by those who are already poor and disadvantaged, making them even more disadvantaged. If we made decisions according to Rawls's notion of justice, we would prioritize saving lives and meeting basic needs, which would require radical GHG emissions reductions and aid from wealthy countries. While bold mitigation actions will cost wealthy countries money (which they have), continued global warming will cost the least advantaged their lives.

The costs are not equivalent, and justice requires focusing on helping poor countries and communities to survive rather than using cost-benefit analysis to rationalize saving money for those who are already the most well-off.

Standing up for justice involves thinking beyond oneself and one's own country to act as a global citizen. According to philosopher Henry Shue, justice in terms of climate change means several things: Mitigation must focus on reducing GHG emissions in wealthy nations, wealthy nations must also assist in "green" development in poor nations to avoid additional GHG emissions, and we cannot ask some people to surrender necessities so that other people can have their luxuries.[15] Allowing people in poor countries to die because helping them would be a bit costly for wealthy countries (who caused the problem) lies at the heart of climate injustice. Therefore, if you live in a wealthy country like I do, we must urge our governments to spend the money necessary to drastically mitigate warming and assist poorer countries in survival, adaptation, and low-carbon development. Rather than thinking of it as an obligation, we can think of it as a standing up for justice, which is an empowering thing to do.

Living the Good Life

Self-help gurus have been selling books for decades about how we can live a good and fulfilling life. What should or shouldn't we be doing with our lives? How do we live well? How do we come to the end of our lives feeling good about our choices and what we have done? Philosophers have also been pondering these very questions for thousands of years. Indeed, many philosophers today still refer to Aristotle (384–322 BC) for guidance. Aristotle deeply contemplated what it means to live a well-lived or flourishing life, or what he called eudaimonia. Eudaimonia is often referred to as happiness, but others contend that it is not a state of being but the process of human flourishing and well-being. It is achieved through virtuous activity or acting with excellence. It is through virtuous actions that we find the path to well-being.

In *Nicomachean Ethics*, Aristotle explains that certain virtues are key to a well-lived life. These include temperance, justice, courage, fortitude, wisdom, prudence, modesty, truthfulness, patience, and friendliness. No one is born virtuous, and it takes practice to cultivate the virtues. Thus we must focus on constructing our character in line with these virtues so that we increasingly behave virtuously in all circumstances as a practice of habit. This is the goal to strive for to have a well-lived, good, or flourishing life. Often referred to as "virtue ethics," this approach to living focuses on the agent, their actions, and their character. Thus when one lies on their deathbed, they know they have lived well by developing a virtuous character and living a virtuous life, and, for Aristotle, this is the "ultimate good" to strive for.

Many contemporaries agree with Aristotle's ideas that living virtuously can increase well-being and foster positive feelings. In cases where one is fighting for something but despite their efforts the desired outcome does not come to fruition, one can still feel good about having acted virtuously despite the results. In other words, with virtue ethics, when the efforts are not successful and the outcome itself is unrewarding, doing what is right and focusing on character can still offer positive feelings. Therefore, following a virtue ethics approach to life can result in enhanced well-being and bring contentment, satisfaction, and serenity. These positive emotions and the personal benefits of virtue ethics make it an attractive way of life. Specific to environmentalism, philosopher Brian Treanor explains that "virtue ethics frames environmentalism in terms of flourishing rather than sacrifice . . . and makes many of the necessary behavior changes attractive rather than onerous."[16]

Many scholars and activists alike have highlighted the importance of virtue ethics as an important motivator for climate activism. Dale Jamieson[17] describes a new ethics for the Anthropocene based on "green virtues" including preservation, temperance, mindfulness, and cooperation. Environmental scientist Mike Hulme[18] suggests that activists focus on the virtues of hope, wisdom, humility, faith, integrity, and love. Lastly, philosopher Byron Williston[19] discusses the importance of three specific virtues for climate activists: justice, truth, and hope. These involve justice for future generations and species, telling the truth about the reality of the situation while rejecting false optimism and technological fantasies, and an active hope for what is still possible. While the lists of specific virtues to focus on may vary, the overall goal is to adhere to these virtues in one's character and actions over the long run.

A virtue ethics approach and the associated positive emotions can help fuel sustained climate activism and in many ways is highly suited to the challenges of the climate crisis. Especially in the past few years, when calls from climate scientists for bold action have been met with little response from governments and when activism sometimes seems ineffective, something more is necessary for individuals to sustain action. As stated by Hulme, "the demandingness of activism needs eudaimonic rationales."

Climate activism is often difficult and demanding, yet knowing it is a virtuous activity and focusing on one's actions rather than the outcomes can help motivate and sustain activism. When conducting interviews with Extinction Rebellion (XR) activists in the United Kingdom in 2019,[20] I found that many held strong beliefs directly in line with a virtue ethics approach. For example, one activist explained, "I have to disassociate from the future. What matters is my actions now and doing the right thing now." Another plainly stated, "In Extinction Rebellion we have to hold outcome lightly. Action is important but the future is beyond us. It's virtue ethics." Doing the right thing is part of living well and drives activists to keep going despite defeats and losses. There are personal benefits of living virtuously, and it also can help the world be a better place. As most of us already seek deeper meaning and want to live a good and fulfilling life, acting virtuously in the context of climate change, especially with so much at stake, makes a lot of sense.

The Joys of Working Together

Working with others, rather than alone, can make a world of difference. Have you ever spent a day doing a tedious, boring, or even disgusting task but ended up having a great time doing it because you were working with other people and enjoyed their company? Then you know this to be true. It is also true for longer-term endeavors and for trying to collectively address social problems. When we work together, we feel group-based emotions that can be based on a shared identity or a shared goal, or both. Having a collective "we" remains critical for sustaining collective action.[21]

The positive feelings that emerge from working together for a common goal can be rewarding, fulfilling, and inspiring. It can also provide a strong sense of belonging and purpose, something many of us seek. When faced with a difficult, unjust, or harmful set of circumstances we can attempt to suffer through it alone or we can come together and try to change these circumstances for the better. Working together for a common goal not only provides a sense of belonging, but it can provide a deep sense of purpose and meaning in our lives.

One example of this was described in depth by the writer and philosopher Albert Camus upon reflecting on his own experience during World War II. Although he was from Algeria, Camus was in France when the Nazis occupied the country and was unable to return home. Living in occupied France, some people went about trying to live their lives normally, attempting to ignore and look away from what was happening to their Jewish countrymen. But others, like Camus, chose a different path. As a part of the French Resistance, Camus wrote articles and other publications promoting the resistance and took part in other underground efforts to save Jewish lives and defy Nazi rule. Community efforts to hide and relocate Jews saved thousands of lives, especially children. As an allegory of these circumstances, Camus's novel *The Plague*[22] emphasizes the positive feelings that emerge when, despite the odds, people come together to help others. While death spread through their community, the main characters worked together to help others and reduce the suffering of those already dying. They experienced strong positive emotions from working together for a just cause. Camus's main argument is that in an absurd and cruel world that seems to have no meaning, the best thing we can do is create meaning by working together to help others.

Camus further detailed the life-affirming emotions associated with working together in *The Rebel*,[23] where he argued that it is human nature to come together to rebel against conditions of injustice and that through solidarity against injustice we can find deeper meaning in life and form connections with others that are otherwise impossible. Camus felt such a strong sense of purpose and belonging working in the underground resistance that when the war was over he desperately sought to find those feelings again. Camus also affirms Aristotelean virtue ethics: he stresses the importance of doing the right thing despite the outcome and having a commitment to wisdom, humility, justice, cooperation, moderation, and sacrifice. Camus's work aimed to motivate action, and it highlights the positive feelings he experienced through virtuously working with others to address injustice and save lives.

Collective action for climate change depends on solidarity, yet solidarity itself has a lot to offer as it can foster positive experiences and emotions. When talking to members of Extinction Rebellion, this emerged as a key factor motivating continued participation. In addition to larger regional groups, XR activists were organized into smaller "affinity" groups that work closely together and look after each other's well-being. Not only did activists work together during sit-ins, protests, and other forms of activism, but they also came together to support one another. They shared grief related to climate change and biodiversity loss. They also forged bonds to help each other physically and to mentally prepare for more difficult times that are likely to emerge on our current warming trajectory. One activist shared that "it feels good to be with people who share something," and another stated "a feeling of belonging is essential to us." Uniting behind a common cause can lead to feelings of rootedness, strength, meaning, joy, and connection and can significantly change our lives for the better.

Building Something New, Something Better

If you take climate change out of the equation, there are still many other social problems. Depending on your situation, you might view some of these as serious problems that need to be addressed: unprecedented economic inequity, corporate power impairing democracies, unemployment, lack of adequate health care, homelessness, hunger, poverty, and debt. In addition, some people lack access to adequate transportation or healthy food, and many communities are plagued with polluted air and water.

Our modern Western society, especially in the United States, is structured in ways that undermine well-being and happiness. Recent statistics from the Commonwealth Fund[24] reveal that, compared to other wealthy countries, the United States has the highest suicide rate and the second highest drug-related death rate. In addition, a quarter of US adults have been diagnosed with anxiety or depression, more than in most other countries. Other studies have solidified a solid linkage between income inequality and mental health problems. For example, there are strong statistically significant positive relationships between depression and income inequality, so much so that mental health professionals state that "policy makers should actively promote actions to reduce income inequality, such as progressive taxation policies and a basic universal income."[25] Mental health is an important indicator of well-being, and in many ways our current system is failing us and needs to be transformed for the better.

In her 2023 book, *The Wealth Supremacy*, Marjorie Kelly clearly lays out the root problems that undermine our ability to address social problems and live in a more flourishing society. Since the 1980s, society has increasingly accepted that we should prioritize wealth accumulation over the health and safety of workers as well as over environmental harms like climate change. Justifications for this priority include obligations to maximize profits for shareholders and the fiduciary duty of financial firms to prioritize profits for their investors. But this "capital bias" puts profits above all else, before people's lives and well-being. Our current economy is biased toward profits, and these profits are mostly going to the wealthy few. We would all benefit from shifting to a more humane and empathetic priority system where other values besides just profit are considered in decision making.

Creating a better society can represent a positive overarching goal that includes addressing climate change. Psychologists Paul Bain and Renata Bongiorno found that the idea of creating a "more benevolent and compassionate society" can be a strong motivator for climate action.[26] They argue that many people are already concerned that society is becoming less caring and moral and that there are fewer community bonds. Many people desire to create a more communal, moral, and caring society. According to their results, this includes addressing climate change, which was widely perceived as a way to help create a better society. The exciting prospects of building something new and better can be a strong motivator for action.

Bold climate mitigation measures can also have a range of other benefits. In other words, in addition to reducing GHG emissions, society benefits from action in other ways. There is ample evidence that mitigation would improve human health, especially health impacts related to air pollution. In addition, Bain and colleagues'[27] study found that people associate additional benefits with climate action, including technological and social development, addressing other societal dysfunctions, and creating a more benevolent and caring society.

As Naomi Klein argues in *This Changes Everything*,[28] addressing climate change involves transforming the very social and economic structures of our society. It represents an opportunity to change society for the better, to improve overall well-being and address other social ills. This is one of the goals behind the idea of a Green New Deal: to address climate change at the same time as economic inequality and other pressing social issues. As philosopher Henry Shue[29] puts it, those who are young now are about to witness something big: either a "glorious triumph" to a better society, or a "dismal failure" where entrenched powers prevent the changes necessary. Let's shoot for the glorious triumph! Climate change offers a real chance to create a better society where people are protected from social and environmental harm and where people are healthier, both physically and mentally.

Since minimizing climate change entails radically transforming our social system, this opens pathways to remake society and to specifically change social policies and structures to increase well-being. Thus, addressing climate change represents a truly incredible opportunity to create a more just, compassionate, and livable society in which we all have a chance to flourish. Being a part of this transformation represents a virtuous as well as an exciting endeavor.

Focusing on the Positive

I am not arguing for blind optimism in the face of climate change, but I do believe that there are many positive motivators for climate action. For years we have been hearing about frightening climate impacts, and public figures keep talking about the potential risks of human extinction if the worst climate trajectories come true. We have also seen the rise of climate doomism and despair, which results in inaction. Yet there is so much that can be done and so much that can be saved. What we need now is more motivation, not less. Focusing on the negative aspects of an impending crisis will only get us so far. We also need positive motivators for actions that reduce global warming, build community, and enhance our personal well-being.

Instead of focusing on fear, guilt, anger, or sacrifice, we have many positive reasons to work together to mitigate climate change. Here we examined how we can focus on protecting today's children, standing up to injustice, living a good and flourishing life, enjoying working with others, and building a new and better society. These are positive, inspirational, and attractive reasons to act on climate change. They are also motivators that can be sustained and that are self-reinforcing. In addition, focusing on these positive aspects of acting makes more sense because no matter what happens, those who act will benefit through living a more connected, meaningful, and positive life. In my view, with all the positive outcomes associated with political engagement in terms of mental health and feeling a sense of meaning and belonging, why wouldn't someone want to get involved?

The next chapter examines what specific "climate action" we need to minimize warming. What should concerned and motivated citizens be doing? What should they demand that their governments do? There are many possible actions one can take, but some will be much more effective than others. Because every fraction of a degree of additional warming matters and negative impacts are not evenly distributed, we must focus on what is most *effective* and *just* when it comes to climate action. While much attention has been placed on the individual to act on climate change, it is through acting collectively that we can have the most influence.

3

What Kind of Climate Action?

Policy should be made on the basis of robust empirical evidence, rather than on the basis of speculative theoretical possibilities, particularly given the severity of the crisis that is at stake.

—Jason Hickel and Giorgos Kallis[1]

On most of the streets in my neighborhood, I can find at least one sign in front of a house that reads "Climate Action Now." It is a striking green and white sign distributed by the League of Conservation Voters. When I look at the sign, I can't help but wonder: What kind of action do they mean? What kind of "climate action" does that neighbor think we need? Is any action good, or are some actions more effective than others? While many people continue to call for "climate action now," it is not clear what this action entails or who should be doing it. Some people might think climate action is putting solar panels on their house or buying an electric car. Others might think it entails supporting policies to increase funding for renewable energy. The truth is that effective and just climate action must entail much, much more.

You Can Reduce Your Personal Emissions, but . . .

There are many reasons to change your personal behavior and consumption patterns to reduce your GHG emissions, but . . . it's not enough. In fact, it is not nearly enough, and I will get into why that is next. Yet even though it is not enough, there are also very good reasons to want to reduce the GHG emissions you are personally responsible for. But if we want to avoid the worst of climate change, we must focus on addressing the bulk of GHG emissions that lay out of an individual's control and require new policies and government intervention.

Despite the propaganda, individual behavioral and consumer-based changes can only take care of a fraction of the emissions reductions necessary. The estimates on the potential contributions of individual behavior changes all confirm that the majority of emissions come from companies and states, not from individuals.[2] For example, a complete shift to green consumerism in the EU (meaning every single person participates, which is unlikely) is estimated to reduce emissions by 25 percent.[3] Another study estimates that the widespread adoption of 30 different behavioral changes could mitigate from 19 to 36% of global emissions.[4] While the initial reduction in GHG emissions due to the Covid-19 pandemic was around 17% one scientist explained that "at the same time, 83% of global emissions are left, which shows how difficult it is to reduce emissions with changes in behaviour. . . . Just behavioural change is not enough."[5] Any changes we do make to reduce GHG emissions as individuals are not enough because the majority of emissions remain out of our reach and therefore require much larger systemic changes.

Individual changes can add up. Most studies show that if everyone does their part to reduce their personal emissions it can add up to a 20–30% reduction in GHG emissions. This is not insignificant. It is difficult, however, to get all individuals to participate in emissions-reducing behaviors in our current system. One problem is that producers are constantly trying to get us to buy more unnecessary things. In contrast to common assumptions, consumption doesn't usually drive production. It has long been recognized that in most cases production drives consumption, as consumer demand is created by the producers or the companies producing the goods. This occurs primarily through marketing campaigns and advertisements. Thus, the producers create demand so they can sell more products. That is why consumption-based campaigns to "change the world" rarely succeed.

In our current system, we are inundated daily with advertising to buy more things, and we have a general lack of information about the GHG emissions associated with all of our actions and consumption choices. Consuming less definitely reduces GHG emissions, yet manufacturers and corporations want us to keep buying more and more, pushing us in the opposite direction of emissions reduction. We will discuss sufficiency policies to address these issues in the next chapter. Here I want to simply point out that it would be difficult to achieve the potential 20–30% reductions from cumulative individual actions because our production and consumption system is designed to keep us always buying more, and this means more emissions.

Additional policies are necessary to make low-carbon living the new norm, and that is something we need to demand that our governments make happen. In addition, we should probably focus on the majority of emissions coming from industry and governments that policies can also address. Since we need to get to net zero, this means almost all emissions need to be cut, and it is most effective to focus efforts on the biggest emitters and the underlying drivers of these emissions. In summary, we need to focus on policies to make low-carbon living the norm and to reduce the majority of emissions that remain out of the individual's reach.

One reason we see such a big focus on personal carbon footprints is that fossil fuel companies and vested interests have worked hard to focus attention on individual emissions in order to divert attention away from policies that would address their own (much larger) contributions. The most cited example is British Petroleum hiring a public relations firm to hype up the importance of individual carbon footprints, releasing a carbon footprint calculator, and encouraging offsetting individual carbon emissions. The company received extensive backlash from environmentalists and climate scientists for their attempt to redirect attention away from their own culpability. We can't let their efforts distract us from focusing on the policy changes necessary to effectively reduce the vast majority of emissions that are essential to phase out.

I try to reduce my personal GHG emissions, but not because I think it is going to significantly reduce global warming. I agree with what one climate activist I interviewed told me: "I like to do individual actions, but I don't kid myself it will change the world." Along with many other people, I like to reduce my GHG emissions for other reasons. Some might say these reasons include "not wanting to be part of the problem" or "trying to do the right thing." These positions have also been examined by scholars who have identified different ethical justifications for reducing one's personal GHG emissions. Let's explore two here.

First, philosopher Paul Knights[6] states that although individual actions are relatively "inconsequential," personal carbon emissions reduction is still "morally obligatory" based on the following argument: "A) To remain a member of a harming group is a moral wrongdoing; B) The performance of consumption actions constitutes remaining a member of a harming group; C) Therefore, the performance of consumption actions is a moral wrongdoing." In other words, if we are living a high-consumptive and carbon-intensive lifestyle, we are members of a harming group, and it is a "moral wrongdoing" to remain a member of this group. We can think of it this way: even if you are only adding a grain of sand to a 100-ton pile of sand, it still makes the pile bigger. In the case of GHG emissions, any additional emissions technically do make the problem a bit worse. Even if it is infinitesimal compared to other sources of emissions, who wants to be part of the problem or knowingly remain a member of a group causing harm to others?

Second, philosophers Christian Baatz and Lieske Voget-Kleschin[7] bring up a different justification for individual actions, even when "inconsequential," based on a notion of moral equity and not exceeding an individual's "fair share" of GHG emission entitlements. This is based on the idea, supported by Peter Singer and others, that allowable GHG emissions should be divided among all people equally. In other words, every person has the same fair share. In this case, an individual's (A's) carbon emissions are morally wrong if: "1. A exceeds her fair share of emissions entitlements, and 2. By emitting, A contributes to a harmful activity." This justification takes a more global perspective and enforces the notion of climate justice. In terms of a carbon budget, why should we be able to make unnecessary or wasteful GHG emissions when other people in poorer countries might need those emissions to survive? Because those of us in wealthy countries have already used more than our fair share of the carbon budget, being a good global citizen means using as little as possible of the remaining carbon budget.

If either of these ethical justifications for limiting your personal GHG emissions resonates with you, you are probably wondering what kinds of personal changes and choices can reduce your contributions the most. Although it is controversial, the most significant thing people can do to reduce emissions is to have one fewer child. Not having a child in the United States has an annual impact 40 times as great as avoiding a transatlantic flight, using renewable energy, not driving cars, or going vegan.[8] This may seem like an easy decision for some people, while others may have a strong desire for a large family. Also, the GHG emissions related to an extra child depends on where that child is born and into what circumstances: children in wealthy families in wealthy countries will have much larger impacts. An article in *Science*[9] empirically confirms that having one fewer child is by far the most significant way to reduce your personal GHG emissions, followed by living car-free, avoiding one round-trip transatlantic flight, buying green energy, and eating a plant-based diet. While much attention is given to acts like recycling and upgrading lightbulbs, these have much smaller impacts than the more significant lifestyle changes listed above. These are personal choices and, in many cases, difficult choices to make and stick to in a society that is not designed to make low-carbon living easy, cheap, or convenient.

I try to be conscious of my GHG emissions, and I encourage anyone who is concerned about climate change to do so as well. For me, it doesn't feel right to be emitting unnecessary and wasteful GHGs that could contribute to harming others. I also believe our social system is not designed to make it easy to cut personal emissions, and in many cases we have to compromise our ethical positions to get where we need to go or to get our jobs done. It isn't easy and sometimes it can make you feel rotten. That is why I strongly believe we need policy changes that make low-carbon living easy and accessible to all. Putting the onus on individuals without systemic changes results in less emissions reductions and more personal frustration, as trying to navigate low-carbon living in a high-carbon society is like trying to swim against a forceful current pushing you in the wrong direction.

Increase Renewable Energy, Energy Efficiency, and . . .

In June 2023, the International Energy Agency reported that "[r]enewable power is on course to shatter more records as countries around the world speed up deployment." As renewable technology costs drop and more policies subsidize renewables, we have seen record-breaking increases in renewable energy production. Yet even if solar power made up half of all *new* energy in a country like the United States, this statistic leaves out some of the most important information about our renewable energy "transition."

An energy transition requires that low- or no-emission sources of energy *replace* carbon-intensive, fossil-fuel-based energy sources. Yet this is not what we see happening. Looking at the data available from national and international energy agencies and institutes makes this clear. For example, the Energy Institute's 2023 report "Statistical Review of World Energy"[10] shows that despite an increase in renewable energy consumption, from 6.5% of the global primary energy used in 2021 to 7.5% of global energy used in 2022 (excluding hydropower), fossil fuel consumption remained at 82% of total energy consumption. This makes sense because overall energy consumption globally increased by 1% from the previous year to a rate 3% higher than 2019 (pre-pandemic) levels. This steady percent of fossil fuel use in a system where total energy consumption increases means more fossil fuels were burned. Therefore, it is not surprising that carbon dioxide emissions also continued to increase, growing 0.8% in 2022. This might not seem like much, but based on their national pledges, countries should be acting in ways that drastically reduce GHG emissions.

At a national level, the US Energy Information Administration reports that "[r]enewable generation surpassed coal and nuclear in the U.S. electric power sector in 2022."[11] Yet at the same time, natural gas increased as a percentage of total energy (from 37 to 39%) and remained the largest source of electricity in the United States. While renewables surpassed coal, reaching over 20% of US electricity generation, coal-fired generation was still 20%. Now, keeping these percentages in mind, add the fact that total US electricity consumption also keeps growing (with the exception of the 2020 pandemic lockdown). As the Energy Information Administration reports:

> Total U.S. electricity consumption in 2022 was about 4.05 trillion kWh, the highest amount recorded and 14 times greater than electricity use in 1950. Total annual U.S. electricity consumption increased in all but 11 years between 1950 and 2022, and 8 of the years with year-over-year decreases occurred after 2007.[12] (2007 = financial crisis)

In other words, unless there is an economic recession or pandemic lockdown, overall energy use keeps going up. They also project that total energy consumption will increase by as much as 15% from 2022 to 2050. When total energy use keeps rising, the percentages of remaining coal and natural gas equal more *total* fossil fuels burned and more *total* GHG emissions. Simply because more renewables are being added doesn't mean emissions are going down. In fact they are still going up due to total increasing energy use.

This is far from new information, as years ago social scientists empirically examined the renewable energy "transition" and found that it was not a transition at all. Sociologist Richard York's work used energy data to show that the general trend resulted in renewable energy additions rather than transitions.[13] Another sociologist, Ryan Thombs, coined the term "renewable energy paradox" to describe this counterintuitive outcome: renewable energy has had little influence on carbon emissions in developed countries.[14] Yet when you look at the data and see increasing total energy use, the paradox makes sense. More renewable energy is only going to reduce emissions if total energy use stays the same or, even better, goes down.

Another paradox relates to energy efficiency. According to the International Energy Agency, "energy efficiency is called the 'first fuel' in clean energy transitions, as it provides some of the quickest and most cost-effective carbon mitigation options while lowering energy bills and strengthening energy security."[15] They report that in 2022, global energy-efficiency-related spending increased by 16% and since 2020 governments worldwide have spent around $250 billion each year on energy efficiency actions. The logic here is clear: If our homes, vehicles, and production processes use less energy because of efficiency measures, then we can reduce total energy use and related GHG emissions. As with the renewable energy "transition," however, the empirical work of Richard York and colleagues shows us that this too is not happening. Energy use still keeps going up despite efficiency gains.

There is a name for this energy efficiency paradox: Jevon's paradox. William Stanley Jevons noticed this same paradoxical outcome looking at coal use in 1865. Some people also call this pattern of increased use after efficiency gains the rebound effect. York and colleagues have shown that at all different scales there are examples of the same phenomena: greater efficiency is associated with increases in resource consumption.[16] This increased consumption can partially or fully offset any energy reductions from efficiency depending on the scenario. They list examples: nations with higher efficiency measures generally have higher rates of growth in energy consumption, state-level energy efficiency policies are not associated with energy conservation, and household efficiency measures are partially offset by increases in energy use.

Why does this happen? In some cases, efficiency triggers increased consumption due to reduced costs. For example, the purchase of a new fuel-efficient car can result in the rationalization to drive more. There are also indirect effects, where savings in one area results in increased consumption and spending in another: saved money from buying less gas rationalizes a trip to Paris. Lastly, in the realm of production, when energy is being saved it can expand new possibilities for manufacturing new goods for the market, which will increase energy use. In this case, companies may rationalize producing more energy-intensive goods when they are saving money through energy efficiency processes. Another example is that energy efficiency standards in cars rationalizes the production of an increasing number of large SUVs and trucks. While energy efficiency may be an important tool to mitigate climate change, the related increases in energy consumption undermine its potential.

In each of these cases, paradoxically, we do not see GHG emissions reduction even as green/clean technologies advance. Growth in total energy consumption reduces the potential gains that these technologies could achieve. Thus, *we need renewable energy, energy efficiency,* and *policies to rein in excess energy consumption.* This also relates to material production, as it requires energy to produce goods. Policies to reduce unnecessary production and consumption can come in a variety of forms and are often called sufficiency policies.

Sufficiency policies focus on living well with enough, not with always more and more. They support increased well-being while reducing excess energy and material use. Even in cases where total energy consumption has started to go down, sufficiency policies would make it go down even more. Given that every degree of warming avoided means lives saved, minimizing GHG emissions in wealthy countries should be the goal. Sufficiency policies include work time reduction, advertising restrictions, and the use of taxes that reduce GHG emissions while also addressing economic inequality. These sufficiency policies will be discussed in more depth in the next chapter. The rest of this chapter focuses on the remaining elephant in the room: fossil fuels and their continued extraction and expansion, despite the threats of climate change.

Directly Phase Out Fossil Fuels

The most basic mitigation strategy that many environmentalists and concerned citizens agree upon is reducing fossil fuel use. If governments truly support a renewable energy transition and mean what they say about reaching net zero by 2050, less fossil fuels should be extracted and used each year, not more. Many governments, including the US government, are not creating policies that reduce fossil fuel use and instead are granting more access to further extraction on public lands. Fossil fuel interests and other wealthy actors have incredible influence over policymakers at all levels of government. Companies, banks, and governments continue to support fossil fuels. Instead, we need direct efforts to reduce fossil fuel use.

This means going beyond market mechanisms that offer indirect and incremental change. Carbon markets—involving an emissions cap, the trading of emissions allocations, and carbon offsets—are favored by industries because they allow more flexibility and options to meet environmental objectives. But the caps have tended to be too low, and both investigative reporters and academic researchers have shown that in many cases, offsets fail to deliver. At this point in time, we don't have the luxury of tinkering with indirect and less effective approaches when it comes to climate change. A carbon tax implemented 10–15 years ago that increased significantly over time could certainly have helped US mitigation efforts. It still may represent a way to nudge producers and consumers toward lower-carbon choices. We need much more than nudges, however. All governments need to go beyond market mechanisms and directly phase out fossil fuels. This process begins with ending fossil fuel subsidies.

The United States and many other countries subsidize fossil fuels, meaning they offer financial support that makes the price of fossil fuels artificially low. This seems counterintuitive for governments promoting a renewable energy transition and subsidizing alternative energy production. Calling for a renewable energy transition while still subsidizing fossil fuels is a contradictory and absurd strategy. Explicit subsidies, used to cut the price of fuels for consumers, have not been decreasing and instead doubled in 2022. In this year, global subsidies equaled $13 million a minute and totaled $7 trillion, according to the International Monetary Fund.[17] This amount equals about 7% of the world's total GDP. While 2022 was a unique year for fossil fuel supplies due to the Russian invasion of Ukraine, fossil fuel subsidies have consistently been significant and are not being reduced or eliminated as promised.

Governments have stated that they will reduce or phase out fossil fuel subsidies for years, and most all have failed to do so. The G20 nations, responsible for at least 80% of all GHG emissions, pledged to phase out "inefficient" fossil fuel subsidies back in 2009 and again at the Glasgow climate summit in 2021. Instead, we see trillions of dollars each year making fossil fuels cheaper and more widely used, which incentivizes further extraction. If subsidies were eliminated, it is estimated that it would cut emissions by 34% by 2030.[18] A report from the World Bank[19] states that "by underpricing fossil fuels, governments not only incentivize overuse, but also perpetuate inefficient polluting technologies and entrench inequality." In 2022, G20 governments gave fossil fuels $1 trillion in subsidies as well as $50 billion in loans from public finance and $322 billion in investments for state-owned enterprises.[20] The countries that provide the most subsidies for fossil fuels include the United States, China, Russia, the European Union, and India.

These subsidies have helped fossil fuel companies bring in record-breaking profits for their shareholders and CEOs. In 2022, five oil companies—Exxon, Chevron, Shell, BP, and TotalEnergies—made $195 billion in profits, their most profitable year in history. *The Guardian* compares this to the $36.9 billion spent a year on the Inflation Reduction Act climate funding.[21] While shareholders make money from fossil fuels, continued increases in GHG emissions limit the possibility that we can stay within global warming targets. Some fossil fuel companies have begun advertising that they are transitioning away from polluting fuels and will be supporting climate-friendly operations, but to date these claims remain primarily a form of greenwashing, with their financial investments failing to match their "going green" rhetoric. Meanwhile, some "sustainable" investment funds—environment, social, and governance (ESG) funds— have also been found to be fraudulent greenwashing as they include investments in fossil fuels rather than sustainable alternatives.[22] The supremacy of profits runs so deep that greenwashing to appease ethical investors makes financial sense.

Instead of phasing out fossil fuels, companies are moving forward with plans for further fossil fuel extraction projects, planning well into the future. Despite global agencies saying that there can be *no new investment* in fossil fuels if we are to reach net zero by 2050,[23] fossil fuel companies and governments are moving ahead to open new extraction sites. A 2022 report[24] finds that almost all oil and gas companies are planning for increased exploration and extraction projects, with $160 billion invested in exploration since 2020, and their short-term expansion plans have increased by 20% since 2021. Another report[25] found that between January 2021 and March 2022 oil and gas companies approved $166 billion for investment in new oil and gas projects, the majority of which are not compatible with staying within 1.5°C. In addition, companies invested more than $35 billion for new projects through 2030, which are all incompatible with staying within 1.7°C, with the majority also incompatible with staying within 2.5°C. Early in 2024, *The Guardian* reported a new eight-figure advertising campaign led by the American Petroleum Institute to bolster public support for further oil and gas extraction.

Governments with pledges to reach net zero by 2050 are also moving forward with new extraction projects. According to Oil Change International's data, the United States alone makes up a third of the planned expansion of oil and gas production by midcentury.[26] They call the United States the "Planet Wrecker in Chief" for leading in new extraction. Fifty-one percent of global expansion is being led by the governments of the United States, Canada, Australia, Norway, and the United Kingdom, some of the countries that are already the most responsible for historic GHG emissions. Oil Change International claims that this is "inexcusable" and that "these countries must not only stop expansion immediately but also move first and fastest to phase out their production and pay their fair share to fund a just global energy transition."

Not only do governments need to stop further extraction, but they also need to phase out many current fossil fuel extraction projects. If countries did stop new oil and gas development, using pre-existing extraction sites would only slow production by 2% a year through 2030 and 5% a year from 2030 to 2050. To keep their net-zero pledges, governments must not only stop new fossil fuel extraction but must also *close about half* of the existing fossil fuel production sites.[27] Instead of investing in more extraction projects, governments should be phasing out fossil fuels and gradually shutting down existing production sites. Governments are moving in the wrong direction and continue to do so.

Examples of governments making bold pledges about climate action and then supporting further fossil fuel extraction include decisions made by leaders in the United States and the United Kingdom. In the United States, the Biden administration approved the Willow oil drilling project in Alaska and also released a plan to sell three offshore oil and gas leases in the Gulf of Mexico between 2025 and 2029.[28] In the United Kingdom, Prime Minister Rishi Sunak approved 100 licenses to drill in the North Sea and indicated that other reserves in Rosebank Field would be approved to support British energy security. Through these decisions, leaders jeopardize global security and the security of their citizens by allowing more extraction, GHG emissions, and global warming.

Lastly, it is critical to understand the role of banks in the fossil fuel expansion juggernaut. Most of these expansion projects would not be possible without funding from some of the world's biggest banks. Many of these banks have made climate-friendly statements and net-zero pledges yet continue to fund fossil fuel exploration and development. These include top US banks JP Morgan Chase, Bank of America, and Citigroup, who gave $3.2 trillion to the fossil fuel industry to expand in the Global South between 2016 and 2022, 20 times as much as is being spent on climate solutions.[29] Again, greenwashing makes banks look like they are on the side of a clean energy transition, yet they continue to fund fossil fuel expansion.[30] Banks continue to support the fossil fuel industry, reducing the chances that we can stay within global climate targets.

Some people fear that ending fossil fuel subsidies and fossil fuel extraction will end up hurting citizens through increased fuel prices and reductions in investments such as retirement funds. A well-planned phaseout of fossil fuel subsidies, however, can increase government funds for social support and climate mitigation. A gradual price increase has been successful in other countries, as has compensating lower-income citizens who might struggle with higher prices.[31] Phasing out subsidies can raise substantial funds for other government projects, and compensation can help address staggering levels of economic inequality.

A study examining the economic impacts of ending fossil fuel extraction and leaving "stranded assets" of fuel reserves found that economic losses would primarily impact the wealthiest 10% of the population.[32] In addition, these wealthy individuals will feel little impact. For example, stranded assets would only impact 1% of total wealth for those with net wealth in the top 1%. In other words, low- and middle-income families need not worry about the financial impacts (e.g., to retirement funds) of stopping new fossil fuel exploration and gradually phasing out half of existing production sites. It is only the wealthiest sector of society that would experience a small decrease in their net wealth due to investment losses.

One more thing about the fossil fuel industry: don't be fooled by their promotion of climate techno-fixes. Increasingly, fossil fuel companies and other vested interests are supporting geoengineering as a solution to the climate crisis. For example, solar radiation management is receiving increased attention and research funding. It involves airplanes depositing aerosols into the atmosphere that reflect light and result in "global dimming," reducing global temperatures. It is touted as a low-cost quick fix to address global warming. But models predict a range of impacts that could include increased drought, crop failure, and famine. In addition, if we were to ever stop depositing aerosols, temperatures would quickly rise, which would result in extreme impacts. This represents a risky approach, yet fossil fuel companies and other vested interests are supporting research and development as it takes pressure off phasing out fossil fuels. While many scientists working on geoengineering solutions stress that mitigation is still the best option, this doesn't stop fossil fuel proponents from promoting techno-fixes that can be used to rationalize the continued use of fossil fuels. Relying on these "solutions" is a risky gamble.

The critical step our governments need to take, and what citizens must demand that they take, is ending fossil fuel subsidies, stopping new oil and gas exploration, and phasing out production from existing extraction sites. While many people support the idea of a "free market," subsidies represent a significant intervention that distorts the market and constrains a transition to renewable energy. Governments must also go beyond using market mechanisms alone to drive an energy transition. In addition to subsidizing and infusing funding into renewable energy, governments need to stop granting any new leases for fossil fuel extraction and phase out access to roughly half of existing sites.

A true energy transition will require direct intervention to tip the scales toward a rapid increase in renewable energy *and* a decrease in fossil fuels. For decades governments have been tipping the scales toward fossil fuels, a choice that is becoming increasingly immoral as climate-related disasters increase. Changing this course requires that citizens demand their governments put the brakes on supporting fossil fuels.

Climate Action Must Focus on Governments

What kind of climate action do we need? More than individual lifestyle and consumption changes. More than increasing renewable energy and energy efficiency. In addition to adding more of these things, governments need to accept the reality that some things need to be taken away. Like wasteful production and consumption of energy and materials. Like subsidies for fossil fuels and granting access to new extraction sites. These all must stop. Yet because fossil fuel interests have incredible influence over governments, this is not going to happen on its own. Citizens will need to demand that governments shift their priorities and put protecting people above financial gain for fossil fuel companies and other vested interests.

While countries also make money from fossil fuel development, this must now be seen as unethical, as it is creating harm that increasingly undermines their citizens' well-being. Framing our governments' backwards and perverse policies in moral terms is a key step toward garnering public support for transformative climate action. Given that the role of government should be to protect citizens, continued support for fossil fuel interests at the expense of public health and safety indicates that citizens need to stand up against this injustice and make their governments work for them, not against them.

All of this may seem like an incredibly difficult sell given the current leadership in the United States and in other countries where there is strong political resistance to climate mitigation. But it is not a demand that will be met overnight. As stated before, it is critical to lay the groundwork for change in advance, to build up public support, to identify the best policies, to get people talking about them and considering them, and to organize and strategize well in advance of the sought-after political opportunity for change. A lack of public policy to mitigate climate change simply means that much more of this work needs to be done.

4

What Else Do We Demand?

Who are the realists and who are the dreamers?
—André Gorz

It might feel liberating to imagine we live in a world free of limitations. Yet in reality this is far from true. Parents impose limits on their children to keep them safe and healthy. They limit their exposure to the sun, their time in front of the TV, and how much junk food they can eat. As adults, we are constantly aware of limits. We limit our personal spending. We limit certain foods and drinks to protect our health. We also limit the time we spend on certain tasks, trying to live a well-balanced life. Beyond personal or social limitations, there are also biophysical limitations in our world. In terms of climate change, we have emitted an excess of GHGs, more than can be absorbed by the Earth's carbon sinks and have therefore crossed a limit. Crossing this limit has and will continue to cause harm, loss, and suffering. Despite this harm, governments have recklessly and knowingly surpassed the biophysical limits that scientists continue to warn them about. Instead of continuing to promote the activities driving GHG emissions, our leaders need to stop, respect limits, and practice moderation.

As mentioned in chapter 2, moderation is a virtue. Greek mythology emphasizes the importance of moderation and respecting limits, especially in terms of nature and the gods. Once a protagonist crosses the line, Nemesis or another god seeks swift retribution. In terms of climate change, our leaders have lost sight of moderation, and now we are all facing the consequences. This problem of excess can only be addressed through moderation and respecting limits. As discussed in the previous chapter, simply adding new technologies to our current system is not enough to limit warming in a system of ever-growing and excessive levels of production and consump-

45

tion. Therefore, we need leaders who will take the necessary steps to reduce not only fossil fuel use but other excesses that push us beyond planetary boundaries. According to leading scientists, as of 2023 we have already driven Earth's systems across six out of nine planetary boundaries, putting ourselves at risk and jeopardizing a future for the human species.[1] It's time to practice moderation and rein in harmful excess before it is too late.

Psychologists explain that one reason climate mitigation strategies face so much opposition is that humans in general tend to focus more on what they might lose than what they might gain, referred to as "loss aversion."[2] In other words, people deeply fear having things taken away from them or losing something they once had. But if we can switch our focus, there truly is much to be gained from acting boldly to minimize global warming. This includes all the personal benefits explained in chapter 2, as well as creating a more livable, just, and sustainable world we can all enjoy. Moderation does, however, require cutting back, and for some people that means less profit and therefore loss. But who should cover these losses?

Ethical principles point to the responsibility of the wealthiest 1% and 10% of society, meaning they should shoulder the burden of losses. Why? Because they have and continue to emit the most GHG emissions (primary contributors), they have benefited the most from the carbon-intensive production system (primary beneficiaries), and they have ample resources to deal with the problem (ability to pay). Therefore, we can design sufficiency policies to rein in excess focused on the richest of the rich, who can handle the losses. Governments continue to prioritize excess wealth accumulation for billionaires, while they knowingly fail to protect the lives and well-being of the global majority. It is time for this to change.

As explained in the previous chapter, infusing funding into renewable energy and energy efficiency remains a constrained approach without additional and more direct government interventions to reduce fossil fuel use. But governments must also implement effective "sufficiency" policies that limit excessive and wasteful material and energy use that undermines mitigation efforts. Without these measures, our other mitigation tools remain constrained and insufficient. Therefore, we must demand sufficiency measures to minimize warming.

Most of these sufficiency policies are not only critical for reducing GHG emissions, but they would also help address staggering levels of economic inequality and related social problems. Focusing on reining in excess and increasing sufficiency is essential for a *just* transition to a low-carbon future. Key sufficiency measures discussed here include curbing excess wealth accumulation, addressing planned obsolescence, advertising restrictions, work time reduction, and transitioning to a climate-friendly food system. These measures, especially if implemented together, would serve to significantly reduce GHG emissions and could also increase quality of life and improve social well-being.

Curbing Excess Wealth Accumulation

While it is not highlighted enough in the media, excess wealth is a key driver of GHG emissions. A 2022 study[3] found that since 1990, the global top 1% of the wealthiest individuals has been responsible for 23% of GHG emissions compared to the bottom 50%, who are responsible for only 16% of all emissions. In addition, while per capita emissions for the top 1% increased since 1990, emissions from low- and middle-income groups within rich countries decreased. A 2023 Oxfam International report found that the richest 1% of humanity is responsible for more carbon emissions than the poorest 66%.[4] Put in different terms, the same study states it would take about 1,500 years for someone in the bottom 99% to produce as much carbon as the richest billionaires do in a year. And the carbon footprint of the super rich 0.1% is 77 times as high as the limit needed for global warming to peak at 1.5°C. How could this be?

Very wealthy people have easy access to GHG-intensive luxuries like private jets, high-end cars, and yachts. They also buy more stuff in general and have bigger homes and often own multiple homes. Also, they tend to travel more. In addition, there are GHG emissions linked to their investments. More wealth and more investments in material production and fossil-fuel-related businesses equals more GHG emissions. This is not sustainable. A 2023 study found that millionaires alone will deplete 72% of the remaining carbon budget to stay within 1.5°C and that the wealthiest 0.01% of all millionaires are responsible for 100 times as many GHG emissions than the wealthiest 10%.[5] In other words, every additional amount of excess wealth means more GHG emissions. The wealthiest of the wealthy are indeed much more responsible for the climate crisis.

Excess wealth is also related to staggering levels of economic inequality. In 2023, Oxfam International[6] reported that the richest 1% of the global population "bagged" nearly twice as much wealth as the rest of the world put together over the past two years. This equals $42 trillion, or two-thirds of all new wealth and double what the remaining 99% of the global population received. In the United States, the middle-income group went from having 32% of total wealth in 1983 to 17% in 2016, while upper-income groups went from having 60% of total wealth in 1983 to 79% in 2016.[7] Since the 2008 recession, the richest 5% of families in the United States have been the only group to gain wealth. News headlines in 2023 revealed that the wealthiest 1% of the population in the United Kingdom is now wealthier than the other 70% combined and that the 50 richest families have more wealth than the combined wealth of half the UK population.

Why has this become the case? Explanations of increasing economic inequality typically include globalization, the decline of labor unions, stagnant working wages, historic racism, land distribution, technological change, and government policies. In addition, investments have made the wealthiest even wealthier, with the top 1% of wealth holders owning roughly half of the stock market.[8] This wealth is typically passed on within families. In the United States, estate taxes (on inheritance) have been shrinking as more and more loopholes are exploited. In addition, capital gains taxes, which is a tax on the increased value of stock assets when they are sold, have also been shrinking as loopholes are increasingly used to avoid paying these taxes. That means more and more wealth is being passed on to wealthy heirs tax-free.

This matters because in addition to resulting in excessive and dangerous levels of GHG emissions, inequality has extremely negative impacts on society. Scientists Kate Pickett and Richard Wilkinson have spent decades studying the social impacts of inequality and have published two foundational books on the topic. They argue that inequality impacts the way people feel, think, and behave. Using empirical data, they show that increased economic inequality in societies is associated with increased anxiety, mental health issues, obesity, teen pregnancy, drug abuse, incarceration, and violent behavior including domestic violence and homicide. It is also associated with negative health impacts and a lower average life expectancy. Pickett and Wilkinson argue that inequality erodes social trust and solidarity, which results in negative mental states, poor health, violence, and other negative social outcomes. In contrast, more equal societies face significantly less of these social problems.

Curbing excess wealth can happen in a variety of ways. The most widely supported means include closing loopholes related to estate and capital gains taxes and creating a range of new taxes for excess wealth. This can include instating new progressive taxes, wealth taxes, an income cap, or luxury taxes on high-end goods. The most appropriate mechanisms can be chosen based on the specific social context, yet the primary goal should focus on curbing excess wealth. Wealth taxes have already been proposed by presidential candidates in the United States, including Bernie Sanders and Elizabeth Warren. Their proposals illustrate how few people would be impacted by a wealth tax: Sanders's plan would only affect people who have more than $32 million and Warren's plan would only affect those with more than $50 million.

Based on their data in 2023, Oxfam estimates that a tax of 5% on the world's richest multimillionaires and billionaires would raise $1.7 trillion a year. These funds could be used for climate mitigation and adaptation, increasing social services and protections, and helping people out of poverty. These should not, however, replace any pre-existing social services or involve reducing any current social protections. As climate change impacts intensify, people will only need more protection and support from their governments. In summary, the richest members of society not only have much more than they need but also emit the most GHG emissions, reduce the well-being of others, and generally degrade social relations. Their influence over government policies can no longer be tolerated, and new policies must be demanded that help societies to flourish and effectively reduce climate-related threats. While curbing excess wealth might be the most difficult sufficiency policy to implement given current political conditions, other policies are likely to gain more immediate support.

Addressing Planned Obsolescence

The low-hanging fruit of sufficiency policies is addressing planned obsolescence, which is the act of purposefully manufacturing short-lived products so that they need to be replaced, thus maintaining higher levels of sales. Planned obsolescence costs consumers more money and wastes energy and materials. Yet it is used to make companies and their shareholders richer. Apple products have an especially bad reputation. For example, due to planned obsolescence, Apple's AirPods are designed to last for only 18–36 months of daily use before they are no longer effective.[9] Others claim that updates on older devices make them slower and unusable. In addition, having different plugs for different ports on every new device requires consumers to keep buying new attachments and plugs to make their devices connect. Despite all this attention, Apple products are far from the only ones where we see planned obsolescence.

In his 2021 book *Less Is More*, Jason Hickel makes a strong case for banning planned obsolescence (and he also supports many of the other policies mentioned here). Hickel also brings up issues with Apple and then explains how household appliances are also made to last for a shorter time than they could so that people must buy new ones. For example, washers, dryers, and dishwashers could be made to last much longer. Instead, what happens is that one small part will break, and the cost of repair is so high that many people will simply opt to buy a new appliance. This keeps appliance companies in business but is incredibly wasteful in terms of the materials and energy used in manufacturing. Appliances could be made to last longer and could also be made to be easily repaired when an issue arises.

Planned obsolescence would not be difficult to address. First, it is already becoming common for cities, states, or even countries to ban single-use products like plastic bags and plastic cutlery. While these are not an example of devious planned obsolescence, they are still products purposefully and openly designed to be used, discarded, and replaced. Already in some European countries, food providers at outdoor public events have a system of using ceramic plates, metal cutlery, and glass or ceramic drinking containers. They are simply collected, washed, and used again at the next event. Banning or restricting products that are made to be thrown away after one use can significantly help reduce material and energy use, and for many of us it makes a dining experience much more enjoyable.

Regulations to restrict planned obsolescence for manufacturing companies can take a variety of forms. Companies could be required to have an extended warranty that involves repairing their products when they break down. In Norway, an extended warranty is required by law: two years for most products and five years for products that are expected to last longer. Another more indirect approach requires products to have labels with their expected lifetime to inform consumers about their options. Specific product types could also have low-carbon standards for manufacturing, which can include longer-lasting products, and durability standards can be put into place for product groups that have known possible lifespans. Companies can also be mandated to provide affordable replacement parts so their products can be easily repaired. In some US states "right to repair" laws have been proposed so that people have access to all the necessary tools and information to repair broken items.

In France, planned obsolescence—defined as deliberately reducing the lifespan of a product to increase the replacement rate—is a criminal offense. The penalty can be up to two years in prison and a €300,000 fine, which can increase depending on the average annual revenue from product turnover. To prove that it was deliberate, though, means that proof of intention must be demonstrated through evidence. Still, some companies have been found guilty. Laws in France also fine producers for failure to provide information on spare parts and to make spare parts available in a timely manner, yet there is no provision about the cost of spare parts.[10]

These policies will surely face resistance from companies, but they can greatly benefit consumers and the environment. Fewer things produced means less energy and resources used, as well as less GHG emission. Additionally, it can save consumers money to not have to replace items as often. As any busy person knows, not having your appliances and technological devices break down as often also saves you time, frustration, and personal energy. It's a win-win in terms of climate mitigation, cost savings, and personal well-being.

Advertising Restrictions

Advertising restrictions are critical for reducing GHG emissions driven by excess production and consumption. Research shows that social norms to encourage low-carbon choices for individuals are effective *only* in the absence of advertising. Juana Castro-Santa and others[11] found that "advertising dominated choice and counteracted the positive effects of the social norm," signifying that "low-carbon norms have a limited effectiveness in changing consumer preferences in a world dominated by advertising." In other words, efforts to reduce emissions remain counteracted by advertising, which continues to strongly influence consumer choices. Advertising restrictions can include banning advertisements for high-carbon products or activities and for luxury and status goods, banning ads targeting children, banning misleading labels, and banning or limiting ads in certain media outlets.[12]

Decades ago, John Kenneth Galbraith identified how advertising plays a key role in creating the desires that fuel increased consumption.[13] Advertising and the media are used to create "false needs" through manipulation.[14] French philosopher Guy Debord called them "pseudo-needs" and explained how they are created specifically to maintain ever growing profit accumulation.[15] Advertising influences how people see themselves and their social status and can convince individuals to buy products to address dissatisfaction. In the context of climate change and environmental issues in general, more products are being marketed as "green," yet these products still increase consumption and related GHG emissions. Buying low-carbon products and "voting with your dollar" support a belief that we can alleviate negative impacts through alternative purchasing, yet increased purchasing still undermines potential mitigation gains.

One area to start with advertising restrictions is restricting advertising directed at children. Even in the United States, where there is scant regulation on market relations, advertisements aimed at children were banned until 1984. Since then, it has been a free-for-all, with commercials on children's television networks making them want to buy the latest new toy or eat the newest sugary cereal. Children are easily persuaded and often relentless when they have decided they want something. These commercials no doubt cause families to spend more money than they want while filling their homes with more unnecessary and unhealthy products. Marketing firms have cleverly designed commercials for children that engage their attention and desires. This manipulation of children is unethical and causes excess production, consumption, waste, and GHG emissions. Many of us parents would especially support a ban on ads that convince our children that they *must* have the newest toy, video game, or junk food item.

Other restrictions could include banning advertising in public spaces and on certain media outlets and web pages. Who isn't tired of seeing billboard after billboard along the interstate? Who isn't tired of pop-up ads on websites when you want to read the news or a weather report? In most cases, these are an inconvenient nuisance that detracts from well-being. Ads online are also based on an invasion of our privacy, using web-based data to target ads in response to our online searches. This doesn't feel helpful; it feels like an intrusion. Yet it is highly effective and gets people to buy things they might otherwise not. This type of targeted advertising, as well as advertising in public spaces, needs to stop. Most people would likely support measures to reduce how much they are bombarded with advertisements that effectively push us to buy more and more unnecessary things.

Advertising is especially pernicious because it reinforces a dominant culture of overconsumption and successfully convinces us how we should think and live. As the famous social theorist Herbert Marcuse[16] stated decades ago, "the mere absence of all advertising and of all indoctrinating media of information and entertainment would plunge the individual into a traumatic void where he would have the chance to wonder and to think, to know himself . . . and his society." We need the freedom and space to think for ourselves and decide what is important to us without manipulation and indoctrination. We need to be able to wonder, think, and know ourselves. We also need to reduce waste and excess to mitigate global warming and ensure a safe future.

Work Time Reduction

Policies for work time reduction (WTR), represent a key lever to reduce GHG emissions and vastly improve social well-being, health, and prosperity. In 1930, the famous economist John Maynard Keynes predicted that by 2030 people would only work fifteen hours a week. He based this prediction on continued increases in productivity, which means the material needs of humans can be met with less and less work required. But instead of reducing production once human needs are satiated, producers have used advertising and other means to create more desires, false needs, and never-ending expansion in terms of bigger homes filled with more stuff, more cars per person, more televisions per person, and so on. In other words, we are working more than necessary to buy things we don't need simply to keep up profits for companies.

In the 1950s, Herbert Marcuse criticized the fact that society forces us to organize our lives around the goal of productive work, something he argued is arbitrary and unnecessary. While humans historically may have needed to work many hours each day to fulfill human needs, due to technology and increasing productivity humans no longer need to work so much to survive. And indeed, in many cases, work is not only unnecessary but is also demeaning and alienating, repressing what it means to be a creative human. With the potential to satisfy all human needs with production levels from decades ago (distribution issues are a separate, yet important matter), why work forty hours or more a week to have unnecessary things when you can work fifteen hours a week and have everything you need, along with the freedom to be creative and the free time for hobbies, art, play, and quality time with friends and family? Marcuse was an early advocate and supporter of WTR, but he was not alone.

Another great thinker, French philosopher André Gorz, pointed out that our system of overworking, overproducing, and overconsuming not only reduces human well-being but is a destructive force in terms of environmental impacts. He argued that "production destroys more than it produces" and "we would live better producing less." Gorz recognized the harms of excess material and energy use and was one of the first proponents of the concept of intergenerational justice, arguing that we must instate production and consumption limits now to protect future generations. Gorz supported the idea of WTR as a key means to reduce environmental impacts, and increasing evidence illustrates that Gorz was right.

Studies show that WTR is associated with significant reductions in

GHG emissions, ecological footprints, and resource use.[17] Significantly shorter working hours can reduce rates of production and reduce pressure on resource and energy use. WTR reduces total energy use, as working hours are associated with energy consumption. Social scientists estimate that if working hours were reduced instead of using productivity gains for increased production, the United States would consume 20% less energy, and if we reduced working hours 0.5% annually over the next century we could "eliminate about one-quarter to one-half, if not more, of any warming."[18] In general, because longer working hours are associated with increased GHG emissions, ecological footprints, and energy use, WTR represents a potentially powerful climate change mitigation strategy. For example, carbon emissions and working hours among US states illustrate a positive relationship, with researchers finding that it represents a key policy lever, or game changer, for climate mitigation efforts.[19]

It is important to recognize that WTR does not necessarily guarantee reduced GHG emissions because leisure time could be spent doing environmentally harmful activities like recreational shopping or travel. But with the implementation of advertising restrictions and wealth taxes, as described above, this possibility is diminished, and other mechanisms can also be used to encourage free time to be spent doing low-carbon activities. In addition, a 2022 study found that WTR along with reducing excess wealth accumulation has even more potential to reduce total GHG emissions than either policy strategy alone.[20]

WTR also has many social benefits. WTR would involve reducing annual working hours to a new standard, without decreases in pay or loss of benefits, and would likely also involve work sharing models. Work sharing allows fewer hours worked while avoiding unemployment. Different ways to implement WTR include limiting the number of working hours per week, increasing holidays each year, increasing sick leave, increasing maternity and paternity leave, and offering pre-retirement WTR transitions. Examples of WTR already exist. Most examples have been temporary policies during economic downturns, but increasingly WTR is occurring in European countries. For example, Germany has reduced working hours, and labor unions are demanding even further reductions. How does this work? It can start with a 40-hour work week and then go down to 36 hours, then 32 hours, and so on.

With WTR we are choosing a society where we all work less, where everyone can find work, and where we all have more time for family and friends, creative expression, and to experience the beauties and joys of existence. We are also choosing a key tool to effectively reduce GHG emissions and ensure a flourishing and sustainable society now and in the future. For these reasons, WTR might appeal to many people. WTR is not only possible but is already happening. Given that wages have remained relatively stagnant compared to GDP, profits, and CEO salaries, wages should not be reduced even as working hours are gradually reduced. As calls for WTR to be included in climate policies increase, it may represent an especially appealing component of a climate platform.

Transitioning to a Climate-Friendly Food System

When designing policies to rein in excess, we cannot forget about the significant role of our food system in contributing to global warming. Based on different estimates, the global food system is responsible for at least a quarter and up to a third of GHG emissions. This includes transportation, refrigeration, and all aspects of the food system. Agriculture itself—the growing of crops and raising of livestock—contributes about 10–15% of global GHG emissions. Additional sufficiency policies are necessary to address the emissions from agriculture. A few key policies could be used to address the biggest GHG emitters. This includes addressing significant amounts of methane emissions from animal agriculture.

It is no secret that meat and dairy production results in significant emissions of the GHG methane. Based on its chemistry, methane is a more powerful GHG than carbon dioxide: over 100 years a methane molecule is 28 times as powerful as a warming agent. This means that a little bit of methane goes a long way in terms of global warming. Livestock are responsible for about half of all food system emissions and 15–16% of total GHG emissions in the form of methane. One cow emits on average 220 pounds of methane each year.[21] As highlighted in the news, some scientists' studies show that feeding cows seaweed can significantly reduce methane emissions. This is great, but it is not enough to solve our cow problem. Where is all the seaweed going to come from? How do we make it the new norm? What about the negative environmental impacts of seaweed farming? The "seaweed solution" is interesting, but it is far from a real and scalable solution. We need to scale back beef and dairy production.

The minimum intervention would be to advocate for and encourage a voluntary transition to a more plant-based diet. Recent studies show that shifting to plant-focused diets in high-income countries could reduce agricultural GHG emissions by 61%. In addition, reducing beef consumption by 90% in the United States and other animal products by half by 2030 would eliminate two billion tons of GHG emissions. Globally phasing out all animal products would add up to about half the GHG reductions necessary to limit warming to 2°C.[22] Based on these estimates, we can safely say that moderating the production and consumption of animal products is a key lever to reduce significant quantities of GHG emissions. Government interventions to encourage plant-based diets could include marketing campaigns, education, and revising nutritional guidelines.

What we are seeing recently, however, is an increase in animal product consumption, with demand globally expected to double by 2050. In the United States, we have seen a rise in the promotion of meat consumption through hyped-up fad diets like Paleo and Atkins and even the glorification of extreme meat consumption, especially among young men. Most meat eaters plan to continue to eat meat, and plant-based eaters represent less than 10% of the global population. Yet the United Nations and leading climate scientists call for significant reductions in meat consumption per capita, especially beef, to reduce GHG emissions. More intervention is necessary to nudge consumers toward plant-based eating. This does not mean people cannot eat animal products, only that new policies are needed to reduce the extent of meat and dairy consumption.

While we do see some universities and other institutions taking the lead on plant-based eating, more is necessary to encourage consumers to buy and eat less meat and dairy. In terms of advertising restrictions discussed earlier in this chapter, this is another example where restrictions on these products could be used to reduce the manufactured demand and fetishization of meat and dairy products. While this is an important step, a bigger step would be to phase out meat and dairy subsidies. The EU and the United States both continue to significantly subsidize and financially support meat and dairy. This market intervention makes the cost of animal products artificially low. According to one 2015 estimate from a UC Berkeley research group, without subsidies a Big Mac that costs $5 would cost $13 and a pound of hamburger would cost $30.[23] The EU and the United States are not alone, as many other countries also subsidize these industries. These subsidies for meat and dairy need to be phased out, as they drive additional GHG emissions. As discussed in terms of fossil fuels, subsidies can be phased out in ways that can protect producers and consumers from significant financial stress.

A 2023 article[24] from Stanford University faculty members examines the current challenges in moving consumers away from meat and dairy toward replacement analogs, like veggie burgers and the newer "Impossible" and "Beyond" products. Analyzing data from both the EU and the United States, they found that current government policies are obstructing a transition toward climate-friendly diets. For example, governments spend most of their agricultural budgets on livestock and feed production systems, have failed to include climate or sustainability information in nutritional guidelines, and have even created regulatory impediments for analog products. Much of this, the authors argue, relates to the power of the meat and dairy sectors and how they consistently lobby not only to preserve or increase the levels of financial support they receive but also to undermine climate and sustainability policies. The authors argue there exists a "core alliance" between policymakers and meat and dairy companies, working together to maintain the status quo. These power relations must be challenged to prioritize social well-being over increasing profits for "Big Ag" companies that control the vast majority of the market.

There are other good reasons to move away from meat and dairy as main components of the human diet. One is known health benefits, which I won't get into here. Another relates to how we are going to feed a growing world population with less and less available land. In his 2022 book *Regenesis*, George Monbiot describes how animal farming takes up 83% of the world's agricultural lands but only provides 18% of human dietary calories. In addition, a plant-based diet reduces agricultural land use by 76% and cuts the GHG emissions from agriculture in half. As more people need food, this land use becomes critical, and consuming a lot of animal products starts to make less sense. A tremendous amount of arable land is used to grow animal feed, an extremely inefficient use of land, as much of the potential calories are lost and that land could be used to directly grow food for human consumption. This land adds up, as grazing uses twice as much land globally as crop production but provides only 1.2% of the protein consumed by humans.[25] In other words, continued high consumption of meat and dairy leads to land uses that are unsustainable in terms of feeding the human population into the future.

Beyond focusing on meat and dairy, there are also other ways to reduce GHG emissions from agriculture. These include reducing excess use of nitrogen fertilizer that results in nitrous oxide gas, another powerful GHG. This can also include managing soils to store carbon and release less nitrous oxide. The transportation sector of the food system is also a significant source of GHG emissions. Efforts to localize and regionalize food supplies could cut much of these emissions. All this requires that governments start transforming food and agricultural policies, including the US Farm Bill and the EU Common Agricultural Policy. Current policies are well-protected by the "core alliance" of large agricultural companies, lobbyists, and policy-makers. But citizens can forcefully demand that these alliances be broken and that policies be remade that prioritize protecting their future.

Less Is More

Jason Hickel's 2020 book is aptly titled *Less Is More* because given the ecological and social threats of our time, less truly is more. Most of our problems can be traced to excess: (1) excess profits, (2) excess wealth, (3) excess production (required for profits and wealth), (4) excess consumption (due to manufactured demand required to make the first three possible), and (5) excess GHG emissions inherently linked to excess production, consumption, and energy use. If we want to minimize global warming, this excess must be reined in through policies that require and encourage *less*. We need fewer billionaires, who contribute the most to GHG emissions and excess consumption, and a wealth tax can help with that. We need less waste caused by broken electronics and appliances that can be addressed by banning planned obsolescence. We need less advertising that creates false needs and drives unnecessary, excessive, and harmful levels of consumption. We need less working time so that we can reduce GHG emissions and improve well-being. Lastly, we need less production and consumption of meat and dairy products and less fertilizer use to create a climate-friendly food system.

All of this *less* will result in *more* things that we truly need. In terms of our personal lives, we can have more free time, more freedom in our thoughts (without so much advertising telling what to think and who to be), more clean air to breathe, and more reliable and long-lasting products that don't break and contribute to waste. As a society, we can have a more livable and sustainable world, where people have their needs met and don't have to face the worst impacts of global warming. Our society can also have more happiness and more well-being brought about through measures that curb economic inequality and result in more harmonious and benevolent social relations. All of this *more* is so much better for us than our governments continuing to prioritize more wealth for the already wealthy through an ever-expanding, wasteful, and reckless economy. There are biophysical limits, and our governments cannot continue to support policies that encourage more energy and material use without crossing these limits.

Sufficiency policies are critically needed in addition to deliberately phasing out fossil fuel use and creating a real transition to a low-carbon society. All of this, however, will not happen on its own, especially with entrenched powers supporting the interests of the wealthy and those vested in fossil fuels. Policy changes that go against these interests must be demanded loudly from a critical mass of concerned citizens. At the same time, work must be done to increase political participation, strengthen our democracies, and increase transparency and accountability in our political systems. However, none of this will happen without collective action.

The Need for Collective Action

Humankind is in a race between two tipping points. The first is when the Earth's ecosystems and the life they contain tip into irreversible collapse due to climate change. The second is when the fight for climate action tips from being just one of many political concerns to becoming a mass social movement. The existential question is, which tipping point will we hit first?

—Eric Beinhocker

In December 2023, more than 1,000 scientists and academics posted a letter[1] to the public urging that citizens become climate activists to demand their governments take bold climate action. The letter states, "World leaders have known about the dangers of the climate crisis for decades, but they are not acting accordingly. It is still possible to turn the tide—but we need you." The path we are currently on, the scientists explained, will cause "untold suffering." Yet there is still time to avoid this future and create a livable alternative. This requires widespread public support for transformative changes. The scientists called on everyone to step up and engage in collective action:

Wherever you are, become a climate advocate or activist. Join or start groups pushing for policies that help secure a better future. Contact groups that are active where you are, find out when they meet and attend their meetings. Find out what kind of engagement suits you best and talk to friends, family, and colleagues to spread the word. If we are to create a livable future, climate action must move from being something that others do to something that we all do.

Collective action is critical to catalyze social change. But despite repeated warnings and rising public concern about climate change, we still do not see the level of collective action necessary to pressure governments to adopt bold climate policies.

In the United States, the Sunrise Movement and Justice Democrats have been a driving force behind climate action. They moved a once fringe issue into the mainstream of the 2020 Democratic presidential primaries and proposed a Green New Deal to fund renewable energy and address economic inequality and injustice. While the candidate they backed, Bernie Sanders, did not go on to the general election, their work caught the attention of the nation as well as the attention of Joe Biden. He invited leaders from these groups to join his climate task force, and ideas from the Green New Deal were infused into the administration's climate agenda. After Joe Biden was elected, the Sunrise Movement continued to press for a Green New Deal, and many of the policies were included in the proposed Build Back Better bill. Unfortunately, due to politicians who remain deeply tied to fossil fuel interests, the bill was not passed by the US Senate. After months of negotiations, a smaller bill was proposed: the Inflation Reduction Act. Despite its flaws, this bill represents the first piece of climate policy in US history and used billions of dollars to boost renewable energy expansion across the nation. This was indeed a big step. But it is not enough. As Sunrise activists know, this is just the beginning, and even in the face of setbacks and losses they continue to regroup, re-strategize, and keep working for bolder climate policies.

While working strategically within the political system is necessary, we also need broader public pressure on leaders to adopt bolder climate policies. Voting for candidates that will take a stronger position on climate is a great strategy, but beyond electoral politics we also need significant pressure on already elected representatives to do the right thing. This requires collective action, as only a critical mass of people can leverage the power necessary to create change. How do we know this? Because it has happened again and again throughout history.

If you have not lived through a time of significant social change, it is easy to forget the critical role of social movements. It should also be noted that those in power don't particularly want people to think about how influential social movements can be. Looking back on US history, the only reason FDR adopted the New Deal was because of tremendous public pressure: people were in the streets demanding change. The driving force behind the civil rights movement was the collective nonviolent actions led by Dr. Martin Luther King and thousands of citizens refusing to accept continued injustice and discrimination. Seeing protestors sprayed with water hoses or beaten on television represented a key tipping point that caused the movement to grow exponentially. When people see that something is morally wrong, many will join in the fight. Another example is the women's suffrage movement. Women were not simply given the right to vote. They had to demand it, take to the streets, and disrupt the status quo until they succeeded.

Social movements, and specifically collective action, are a critical component of social change. People gain power and leverage by coming together and refusing to participate in the system that is harming them. Movements tend to arise when people's needs are not being met. Unfortunately, with climate change, we need to act in advance of the climate catastrophes that will cause harm and loss to millions of people. We need to act before our needs to be protected are not being met. In addition, most successful social movements have been focused on granting new rights to certain people, ending wars, or ousting a leader from power. All these are specific demands that are much more straightforward. Preventing significant climate harm is much less specific and much more vague and intangible. These aspects of the climate crisis present significant challenges to the climate movement.

These challenges, however, are not insurmountable. Yet they must be identified and overcome through specific strategies. We can learn from the scholars and activists who have studied social movements in the past to learn what the "essential ingredients" for success might include. According to the late sociologist Erik Olin Wright,[2] we must first identify and confront those maintaining the status quo. In this case, we have ample evidence that fossil fuel companies, utilities, and others vested in fossil fuels are actively working hard to maintain the current fossil-fuel-dependent system. More efforts need to publicly expose these efforts and erode the legitimacy of those who support them. This means highlighting the immorality of our leaders who support vested interests instead of their citizens. Wright argues that framing an issue in moral terms can have the greatest impact, for example focusing on harm to children and those who have contributed the least to the climate crisis. Lastly, Wright says we need to simultaneously erode the old policies and programs that don't work and put in place new ones that do. Usually this takes time, but we don't have much time when it comes to the climate crisis. That is where "the logic of rupture" comes into play. Wright argues that collective actions that are disruptive to the system can pressure leaders to take more urgent and bold action.

Frances Fox Piven, a professor of political science and sociology, provides some additional insights as to how we can speed up the process of social change.[3] Piven's work highlights how power can be harnessed from everyday citizens to challenge ruling parties. How is this done? It all depends on the collective aspect of collective action. When masses of people refuse to cooperate in the system and can bring it to a halt, they can successfully challenge power from above and force leaders to yield to their demands. This makes sense if we remember the legendary force of labor unions who rose up in solidarity to strike against injustice in the workplace. This also explains the decades of efforts from conservative political figures and elites to destroy labor unions and undermine collective action altogether. Amassing the large numbers of people to leverage power is difficult, but it can be highly effective.

The good news is that the climate movement is larger and more influential than ever before. The bad news is it is still far away from being able to challenge power. But there are a range of new tactics being used and more people are joining the movement every year. Let's briefly review what the movement has achieved thus far.

The Climate Movement

One of the first glimmers of an emerging climate movement was the 2014 Peoples Climate March in New York City. This widely publicized event, however, was not followed up with sustained large-scale actions that would raise support for further participation in the movement. More recently a larger and sustained climate movement has emerged. Since 2018, the climate movement has grown to an unprecedented size, with many large-scale global protests demanding government action. Multiple climate movement organizations have taken the spotlight using a variety of tactics and strategies, all of which have served key purposes.

I already mentioned the Sunrise Movement in the United States. They have diverse strategies. They work within electoral politics to (1) get people who support bold climate action elected, (2) get elected officials to pledge to support bold climate action, and (3) work with think tanks and elected representatives on specific policy ideas, like the Green New Deal. In addition to these strategies that directly engage in politics, they also use disruption as a key strategy to raise awareness and directly pressure leaders to act. For example, they famously held a sit-in in the office of Nancy Pelosi, the then Speaker of the House. They have also followed representatives around to pressure them to act, including when Senate Democrat Joe Manchin refused to support the Build Back Better bill. Many of these are acts of civil disobedience, meaning they involve breaking the law (although peacefully) and can lead to arrest. This dual approach serves to infuse new people and ideas into politics and put outside pressure on leaders to support bold climate action. Despite clear setbacks resulting from the November 2024 US election, Sunrise continues to strategize and plan ahead for opportunities for future climate policies and has no plans to give up.

Fridays for Future (FFF) was inspired by Greta Thunberg, who at 15 was protesting alone outside the Swedish Parliament every Friday, demanding action to address the climate and ecological crises. This resulted in the youth-led group Fridays for Future carrying out Friday school strikes for the climate. FFF gained increased attention in 2019 when approximately 1.5 million students participated in a global strike in March and around 6 million people participated in two consecutive general strikes in September.[4] These strikes and those that continue globally have been key to increasing awareness that children and young people are very concerned about climate change jeopardizing their future. They are not necessarily disruptive, as many are approved by law enforcement in advance. Still, FFF represents a powerful force to frame climate change in moral terms, bringing attention to the many children whose futures will be negatively impacted by the climate crisis. As FFF and other youth-based climate groups put more attention on the immorality of the status quo, more people may support the movement.

In 2018, Extinction Rebellion (XR) initiated a wave of protests and acts of civil disobedience in the United Kingdom that spread internationally. Their strategy from 2018 to 2023 was to repeatedly shut down centers of government and commerce (e.g., London in the United Kingdom) until their three demands are met. They demand that governments (1) tell the truth about the climate and ecological crises, (2) reduce carbon emissions to net zero by 2025, and (3) create a democratic citizens assembly to guide the policies for this transition. Some of their rebellions did successfully shut down key areas of London and other cities for up to two weeks, simply through thousands of people sitting peacefully and refusing to move from key routes. For example, in November 2018, more than 6,000 activists shut down five major bridges in London. XR continued to shut down central areas of London and other cities in the United Kingdom during annual fall and spring "rebellions." These events resulted in thousands of activists being arrested for civil disobedience (in this case illegally taking up space or chaining or gluing themselves to property). They also caused massive disruptions and angered many people. Even though many people were angry at XR activists, after each of their rebellions in London, public concern about climate change increased in the United Kingdom. Meaning they were successfully raising awareness.

XR, however, has had limited success getting governments to meet their demands. Their demands are vague enough that they have been easily co-opted in ways that governments can say they are meeting the demands. For example, the United Kingdom declared a climate emergency, but this did not entail bold mitigation actions along with the declaration. The UK government also held a citizens assembly to hear ideas about how to address the climate crisis. But the input from citizens—which called for bold actions—was not transformed into policies. Lastly, although the United Kingdom is one of the leading nations in terms of reducing national carbon emissions, their goal is to reach net zero by 2050, not XR's goal of 2025. Some have criticized XR's strategy of staying out of politics, saying they lack a concrete policy agenda to push forward. This may have resulted in the disappointing responses to their three demands. More recently, news articles in the United Kingdom indicate that XR might be moving away from tactics of disruption and more toward communication and working with governments to find solutions.

One reason XR might be pulling back on disruptive tactics is that other groups have emerged that are increasingly disruptive. This includes Just Stop Oil. Just Stop Oil has become famous (or infamous) for its seemingly random acts of disturbance and civil disobedience, which have included blocking fossil fuel transport and acts of vandalism. For example, Just Stop Oil has blocked roadways and vehicles, disrupting the transport of fuel to major cities. Just Stop Oil activists also made headlines for throwing a can of soup at a famous Van Gogh painting (the painting was not harmed due to a protective barrier). While some people criticized this move as seemingly unrelated to climate change and pointless, Just Stop Oil leaders claim it served an important purpose: to bring attention to their group and the need to stop using fossil fuels. Activists have also disrupted sporting events and other public gatherings to gain attention. Is all attention good attention? This question is still out for debate, but maybe there is an important role for more disruptive groups like Just Stop Oil.

More recently, those who study the climate movement, such as sociologist Dana Fisher, have been talking about the importance of a "radical flank" in the climate movement. The radical flank is a part of a movement that is willing to take more risks, be more disruptive, break laws, and generally take more drastic actions. History shows that many social movements have had a radical flank. This includes the Black Panthers in the civil rights movement and the groups of suffragettes who bombed mailboxes in London. Some argue that these fringe groups serve to increase attention (often in a negative way) and can help the main core of the movement that is not radical to gain more support. In this case, groups like Just Stop Oil might result in initial backlash but then later more support for more mainstream climate groups like FFF or XR. Currently we are seeing increasing escalation between climate groups and law enforcement, as well as angry citizens rising up against climate activists. As this escalation increases, we might see events where some climate activists are harmed or treated unjustly by authorities. This historically has led to more support for a cause, yet it can put the safety of activists at risk. This escalation, however, may be inevitable as the stakes rise for both sides. As activists are taking bolder action, we see more cases of angry citizens rising up in opposition and more pressure from fossil fuel interests for governments to arrest and punish activists.

More and more laws are being passed at state and national levels that prohibit certain types of protest and activism and create harsher penalties for breaking these laws. Some call for extensive jail time for disruptive actions like blocking roads and blocking fossil fuel infrastructure development or transportation. Fossil fuel companies have a strong incentive to push these laws forward and impose harsh consequences for disruptions to their production and transportation operations. They aim to make people less willing to participate in disruptive actions because the consequences are so severe. Critics of these antiprotest laws claim that they are unconstitutional and undermine a key pillar of democracy. If we don't have the right to protest, how can we ever challenge our governments' actions? Activist groups must now be increasingly careful about which strategies they choose. With more governments adopting policies that criminalize certain kinds of protest and activism, some groups may be considering more covert strategies that include sabotage.

Andreas Malm's book *How to Blow Up a Pipeline*, and the resulting film of the same name, raise the question of how "radical" the radical flank should be. Taking up space, running across a professional soccer field midplay, or throwing cans of soup at paintings are one thing, whereas destroying fossil fuel infrastructure is seen as an act of violence, even if the aim is property not people. Some think this is a necessary next step for the radical flank of the climate movement and might be the only way to challenge power. Others argue it will undermine broader public support and backfire, much like the violent actions of animal rights activists in the 1980s.

Peter Singer has since referred to these acts as the major mistake the animal rights movement made that lost them what could have been much broader support. Whether or not we see acts of property destruction in the near future might depend on how governments respond to current pressure, and if they don't, how desperate activists feel. Other scholars[5] have argued that nonviolent civil disobedience, breaking the law through peaceful means, is much more effective for building a larger movement and getting governments to meet citizens' demands.

Another climate movement tactic that can serve to blame and shame bad actors is suing. More than 20 US cities and states have filed lawsuits against fossil fuel companies for climate-related damages or for lying about their product's harmful impacts. An environmental coalition in the Netherlands sued Royal Dutch Shell for violating human rights, and the district court in the Hague sided with the environmentalist, ordering Shell to reduce emissions by at least 45% by 2030. Shell has appealed this verdict, and it will likely be reviewed by an appeals court. These cases serve to publicly blame fossil fuel companies for their actions and help to shine a light on the immorality of their knowingly selling a product that causes harm.

In addition, young citizens have filed lawsuits against their governments for failing to prevent climate harm. In the United States, a case made national news when a Montana judge surprisingly sided with youth plaintiffs. The judge ruled that the state was violating their right to a "clean and healthy environment" and that approving fossil fuel projects in the face of climate change is unconstitutional. This ruling was upheld by the Montana Supreme Court in 2025, yet state representatives are attempting to meet the conditions of the ruling without negatively impacting the fossil fuel sector. Another high-profile case is *Juliana v. United States*, where youth plaintiffs sued the US government for violating their rights to life, liberty, and property. After decades of national attention, the case ended in 2025 when the Supreme Court refused to revisit a lower court's decision to dismiss the case. Despite the outcome, the case brought considerable attention to the many young people who are demanding a safer and better future. At least another 20 youth climate lawsuits have been filed across the globe, including in Austria, Indonesia, Quebec, Australia, Ecuador, Brazil, Germany, and the United Kingdom. Although these trials keep being drawn out through continued appeals and are sometimes dismissed, the lawsuits do successfully bring public attention to the injustices of climate change and the moral obligation of governments to protect young people from climate harm. These tactics might not be viewed as activism, but they can effectively communicate climate concern and help frame the climate crises as a moral issue.

Strategies Moving Forward

Given what has already been accomplished and that so much still needs to be done to curb GHG emissions, I will highlight three areas of strategy that could give the climate movement more power, leverage, and impact. These include increasing the extent of participation in climate activism and political engagement, supporting more specific "game changing" policies, and continuing to focus on the immorality of climate inaction and the resulting harm, especially to children and young people.

As we learned from Frances Fox Piven, challenging power requires having a critical mass of people working in solidarity for a cause. While climate protests and school strikes raise awareness and can build support for the climate movement, it will take a greater number of people to push bold climate actions forward. For example, to get the US government to stop subsidizing fossil fuels, it will take persistent pressure and possibly some amount of system disruption. Governments will need to be pressured into switching their alliances to protect people over profits, and this will require significant participation from a broader base of citizens.

Will increasing public pressure be enough or will the climate movement need to more directly disrupt the system to challenge and leverage power? Piven argues that leveraging power must involve a critical mass of people refusing to cooperate with the system. As Piven explains, "leverage can in principle be activated by all parties to social relations, and it can also be activated from below, by the withdrawal of contributions to social cooperation." This can involve widespread strikes, peaceful sit-ins, or disrupting transportation or finance. Yet this leverage depends on solidarity, a united front involving a significant portion of the population.

Even though the climate movement is larger than ever before, more people will need to participate to challenge power. How many more people? Many scholars and activists refer to the book *Why Civil Resistance Works* by Erica Chenoweth and Maria Stephan, who claim that when a critical mass of 3.5% of the population participates in sustained nonviolent disruption, it can trigger large-scale social transformation. Their historical research illustrates that nonviolent civil disobedience is more effective than violent rebellion, and in most cases movements have been successful at pressuring government action if the participation in civil disobedience crosses the 3.5% threshold. For the climate movement, this would require a significant increase in participation. For example, in the United States this would require about 12 million people to participate. In contrast, the 2019 global strike, with coordinated protests on all continents, had about 6 million participants, which was only 0.07% of the world population. In other words, many more people need to get involved to have a more powerful impact.

How do we increase support for and participation in a mass movement to put people's lives before fossil fuel interests and wealth accumulation for the few? This might require what some people call a "movement of moments," where all groups fighting for climate justice *and* social justice come together to fight for a better future. This involves finding commonalities across different movements and focusing on intersectional interests including addressing inequality, racism, and other forms of oppression. Sociologist John Bellamy Foster[6] calls for a "planetary revolt of humanity" and argues that this might be led by groups in the Global South who increasingly face the worst climate impacts. Movements tend to grow when an increasing number of people's needs are not being met. For many in the Global South, this is already occurring and will only increase. Building solidarity will require common goals and coordinated actions. Focusing on the national level makes sense because most climate policies are needed at the federal level. But globally coordinated actions are also important to show world leaders that all affected people need protection from climate change.

Second, climate movement groups might have more success pushing forward bold climate policy if they get specific about what they want and refuse to start by proposing an already compromised platform. We have already seen how many policies to address social problems get watered down and weakened through political negotiations and compromises. Instead of starting with a compromised proposal, it is better to start with bold actions and try to keep them from being undermined or weakened through political negotiations. For example, demanding that governments end fossil fuel subsidies is a key demand. But what about demanding that our government end fossil fuel subsidies and phase out half of existing extraction sites, which is necessary to stay within global climate targets. Then perhaps a compromised or weakened policy might at least include ending fossil fuel subsidies. We have seen what happens when progressive politicians begin with a middle-ground proposal, and with climate change we can't risk having only partial or weak solutions implemented.

In addition to going bold, groups like the Sunrise Movement and others who work in the realm of climate politics need to start demanding specific policies that not only increase renewable energy, energy efficiency, and green jobs, but that also start to rein in excess. Demanding a wealth tax would face strong opposition, yet demanding it publicly would begin the process of getting more people familiar with the idea and how it would work to address GHG emissions and economic inequality. Work time reduction is something else that should be included in climate policy proposals like a Green New Deal. Like wealth taxes, it is a win-win way to reduce emissions and increase social well-being. In fact, all the policy options discussed in chapter 4 could be proposed together to create a much bolder and more transformational Green New Deal or similar policy platform. To maximize the effectiveness of our mitigation tools and to minimize global warming, we must address excess through sufficiency policies. Otherwise, they will undermine our efforts. This needs to be communicated more widely so that our next round of climate policies addresses these issues rather than simply infusing money into renewable energy and energy efficiency. These are "game-changing" policies that can help to effectively reduce GHG emissions and transform our society for the better.

Lastly, further emphasizing the moral dimensions of the climate crisis is likely to increase support for climate policy and participation in climate activism. Erik Olin Wright[7] stated that most people are motivated by moral concerns, even more than economic or class issues. In his article "I am a Carbon Abolitionist," economist Eric Beinhocker[8] put it simply: "the future of our species and planet depends on creating a mass social movement motivated by moral arguments, not statistics." When people feel that something is morally wrong, they are much more likely to act. Figures like Greta Thunberg have already helped frame climate change in moral terms. Youth strikes also have been key in bringing more attention to the fact that young people are being robbed of a future free from climate-related harm, loss, and suffering. More youth lawsuits against governments can also help to direct more attention toward the immorality of inaction to protect citizens. As Beinhocker argues in the same article, winning social movements have "moral arguments at their core: the way things are is wrong and must change."

It would be unfortunate if it took a massive climate-related disaster with many fatalities to get people to see climate change in moral terms. Yet these types of events are likely as governments continue to fail to act boldly to minimize global warming. This needs to be seen as immoral on their part and as a failure. Moral framing can help delegitimize leaders who fail to act and get more people elected who will act to minimize warming. Moral framing is likely the biggest key factor in getting broader support for climate action. We need to see climate change not as a political issue, but as a human issue, focusing on saving lives and protecting children. This could represent a key tipping point that could help build a critical mass of participation in activism as well as more political leaders and voters supporting action.

We also need to acknowledge the role of social tipping points in transformative change. A 2020 article[9] authored by 14 leading scientists stresses the importance of social tipping points in the context of responding to climate change. They identify six possible areas for social tipping points to emerge in ways that could catalyze bold responses to the climate crisis. Among the possible tipping points is the decentralization of energy systems, which would allow for a more rapid transformation toward renewables, the potential of city climate policies to spill over and be adopted in other places, as well as shifts in knowledge and access to information. Also included and discussed at length are changes in norms and values. The scientists explain the importance of a shift in perception about the morality of continued fossil fuel use. Citing the abolition of slavery in the United Kingdom, the authors explain how changes in ethical perceptions, instigated by a small group of intellectuals, spread widely and resulted in a massive social change. As more people saw slavery as immoral, more people supported abolition. In terms of the climate crisis, the researchers suggest that groups like Fridays for Future, Extinction Rebellion, and the Sunrise Movement are already making headway in terms of exposing the immorality of inaction. At what point will the majority see continued emissions as immoral? What can we do in the meantime to build public support for the necessary policies and help catalyze a social tipping point?

What Should You Do?

The climate movement has come a long way but needs to go much further. This momentum already exists and needs to be pushed forward with increasing levels of public and political support. Here we have looked at a few specific strategies that might help increase power and influence. There are many more to be considered as well. What we do know is that we need more people involved. And that means all of us.

Given all this information, what should we as individuals do to strengthen and further efforts to demand bold climate action? What should *you* specifically do? Philosopher and activist Elizabeth Cripps[10] provides some advice. Cripps suggests that we all consider what our strengths and abilities are and find ways to get active or politically engaged that don't undermine our own well-being. Some people have devoted their careers and lives to working for climate justice, but many people may need a more balanced approach to make climate action fit into the commitments they already have. Finding your strengths is great advice. This can range from simply talking about climate change more, communicating policy solutions, participating in civil disobedience, or running for office. All of this helps, as does increasing voter turnout and working on specific pro-climate campaigns. Whatever your skill set is, you can put it to use. Not all of us can be full-time activists or politicians. And that's okay. We need all types of actions taking place in all parts of society. Everything counts. As Britt Wray explains:

> Stumbling out of one's daze of comfort into an existential battle for the future of humanity isn't something many people feel equipped for. Often, thoughts of helplessness, such as *What can I do? I'm only one person and this is all so baked in,* can be immobilizing but actually, any pro-caring, pro-future, pro-environmental actions will help, and there are endless ones to choose from.

While there are things we can do on our own, collective action is necessary. As world famous climate activist Bill McKibben consistently tells people, the most important thing you can do to address climate change is to join a group of other people and work on it together. We cannot do this alone.

6

Changing Our Minds,
Building Something Better

Progress is impossible without change; and those who cannot change their minds cannot change anything.

—George Bernard Shaw

To build the public support necessary to pressure governments to adopt transformative climate policy, it is going to take some work in terms of changing the way people think. This chapter first discusses public opinion and ideas about climate change action broadly and then what types of changes in accepted ideas and beliefs might be necessary to adopt some of the more transformative policies outlined in chapter 4. These are not only policies that have high potential to minimize warming, but they can also help us transition to a more just and sustainable society. All this, however, requires shifting the way people think about society.

First, we need to address the primary question: How do we garner enough public support for bold climate policy? The surprising news is we already have enough public support, but most people don't know it. Second, how do we garner enough public support for specific policies that prioritize equity, sustainability, and ecological and human well-being? This requires some deeper shifts in our thinking in terms of the dominant ideologies that shape what we think is good/bad, desirable/undesirable, natural/unnatural, and possible/impossible. For decades our ideas and beliefs have been purposely shaped by those with an agenda to maximize profits, but if we stop and think about some of these beliefs and where they come from, they start to lose their appeal. In order to adopt transformative policies that can justly minimize warming, some widely accepted ideas need to be

73

challenged. How do we go against the ideological tide and transition to a society that prioritizes sustainability and well-being? These are difficult questions to answer, but thinking critically about the role of ideas is essential for social change.

Changing Our Minds About (Support for) Climate Change Action

Widespread vocal support for bold climate action is hindered by some fascinating psychological phenomena. While most people assume support for policies to address global warming is low, it is in fact much higher. But believing you are in the minority in terms of support for climate action, people tend to think that there are not enough people for a climate movement, that there is no way a bold climate agenda will get enough support, and that you probably shouldn't talk about climate change to most people because they won't agree with you. This leads to inaction and, worst of all, silence about climate change.

The surprisingly good news is that most people already support climate action. As mentioned in chapter 2, the majority of US citizens and citizens polled across the globe want governments to act to address climate change. Many assume, however, that support for climate action is much lower than it is and are therefore less likely to talk about it or join a social movement. Remember the 2022 study[1] finding that 80 to 90% of Americans underestimate public concern about climate change and support for bold climate policies. In fact, a supermajority of US citizens (66 to 80% depending on the policy) support climate action. Yet the dominant belief is there is only about 3% of the population who think this way. And remember that a 2023 study[2] found that in 23 different countries, 77% of people agreed with the statement "It is essential that our government does whatever it takes to limit the effects of climate change." This is indeed a supermajority who assume they are not the majority. If more people knew this, they would be more likely to join a climate movement organization or at least talk to other people more about climate change and what should be done.

Cass Sunstein, in his book *How Change Happens*,[3] argues that to create cascades of social change we have to break down the silence surrounding an issue and have a better sense of what the majority of people actually think. Breaking away from the "spiral of silence"[4] that hinders talking about climate change and opening up diverse deliberative conversations is key. When people falsely assume that they are the only ones who believe something, it is what was referred to earlier as "pluralistic ignorance." This false assumption results in silence about the topic and can also result in what Sunstein calls "preferences falsification," which is going along with what you think is the dominant belief despite actually disagreeing with it. When one assumes everyone else agrees with the status quo, it silences those who want change. In some cases, like climate change, this is the majority of people. But when a few people start to vocalize concerns, such as in the MeToo movement, others start to speak up, revealing that masses of people want change. While we might assume people don't want bold climate action, the polls reveal otherwise. These misunderstandings need correcting, and more people need to express their true opinions about climate change.

In *How Minds Change*,[5] David McRaney explains that changes in public opinion can sometimes be rapid and happen surprisingly fast. Even when there have been no signs of meaningful progress for years and it seems as if the status quo is how things will always be, an unexpected event or the random start of a cascade of change in social dialogue can trigger a shift in public opinion. Starting with the early adopters and ending with the few remaining "holdouts," people begin to accept new ideas through cascades or chains of interactions and conversations. McRaney cites examples of rapid change in public opinion such as same-sex marriage and the legalization of marijuana, where there were indeed "rapid, surprising, status-quo-shattering global cascades of change." Again, many people might support a change but assume no one else does. But once people realize they are not alone, it becomes easier to talk about an opinion, and once a threshold has been breached, shifts in the way people think about an issue can spread like wildfire.

McRaney also explains that for remaining holdouts, many have had their minds changed about an issue through "deep canvassing," which involves asking people about their views, asking why they think this way, and giving them space to question and analyze their own perspectives. This turns out to be much more effective than presenting information and statistics on an issue or trying to convince someone to change their opinion. This strategy worked impressively well for those who opposed same-sex marriage. People don't want to be told to think differently; they need space to examine and understand why they support a certain position, and sometimes they find that it is not for very good reasons.

In summary, majority support for governments to act boldly on climate change exists despite the holdouts who remain extremely vocal about their opposition. But because climate change has been framed as such a divisive issue, people don't talk about it nearly as much as they think about it.

Doomerism plays a role here too. Why would you want to bring up a topic that can make people feel scared or depressed about the future? Yet we are not doomed (see chapter 1) and, as laid out in chapter 2, there many positive reasons to get involved in climate activism. But acknowledging the bad parts and *all* the emotions associated with climate change is also necessary. As Britt Wray writes in *Generation Dread*,[6] activism must include not only external work, but internal work to recognize the range of emotions being felt. For Wray the ultimate questions are: "Are we going to let our feelings overrun and deplete us, or are we going to use our feelings to overrun the systems that are making us so unwell?"

To change the system driving us deeper into climate disaster, we need to express these feelings, share our opinions, and realize that we are part of a supermajority that wants governments to adopt bold climate action. But we must also recognize that in the United States our democratic processes have been weakened by election manipulation and executive overreach, and our elected representatives in many cases do not vote in ways that reflect public opinion. Being in the supermajority isn't enough if elected representatives don't act accordingly. That is why collective action that brings attention to these issues is so critical as well as voting out incumbents that fail to represent their constituents.

Why Ideology Matters

Beyond increasing and vocalizing general support for governments to act boldly on climate change, we need to increase public acceptance of specific policies and strategies that would most effectively and justly minimize warming. Policies must have broad support if they are going to be adopted, especially ones that go against the interests of fossil fuel companies and others vested in a high-carbon system. Wide support for the policies discussed in chapter 4 will take not only broad support for government action, but it will also require different ways of thinking. Dominant beliefs currently constrain what is possible in the realm of climate change mitigation. Beliefs and sets of beliefs (or worldviews) are also referred to as ideology. Most people associate the word *ideology* with political ideology, for example being a conservative or a liberal or on the right or the left. But there are a range of ideologies. Over time, dominant ideologies change and allow for new types of governance and policies. For example, dominant views in the United States about equality and rights (e.g., slavery, civil rights, gender equality, and sexual orientation) changed over the course of history, which resulted in laws that represent these changes.

The work of sociologist Göran Therborn[7] can help us better understand why ideology matters so much in terms of addressing climate change. According to Therborn, ideologies are not constant but are always in flux. They direct how people make sense of the world and how they live their lives. He explains that ideology manifests in how individuals recognize (1) what exists, (2) what is good/bad or right/wrong, and (3) what is possible. You can likely already identify differences between what a right-wing conservative would think versus a left-wing liberal about what exists, what is good or bad, and what is possible in terms of climate change. Sometimes our ideologies remain hidden assumptions that we don't even fully recognize we have. Dominant ideologies change, but often not easily and not overnight. Therborn explains:

> In order to become committed to changing something, one must first get to know that it exists, then make up one's mind whether it is good that it exists. And before deciding to do something about a bad state of affairs, one must first be convinced that there is some chance of actually changing it.

In terms of climate change, we have to first agree it is happening, then decide it is not good or desirable for it to continue to happen, agree that we should do something about it, and believe that there is actually a possibility that we can do something about it. Views on climate change have shifted over time: more people agree that climate change is happening and something should be done about it, yet dominant ideologies constrain ideas about what we should do about it and what is possible to do.

Contrasting ideologies result in people having different beliefs about what is good/bad, right/wrong, possible/impossible when it comes to climate change policies that could help minimize global warming. For example, if someone thinks government intervention is inherently bad, they are unlikely to support the policies discussed in this book to reduce GHG emissions. If one believes that growth is always a good thing, they are unlikely to support reducing production and consumption to mitigate climate change. Thus, to garner enough support for bold climate policies that have the potential to minimize warming, we must address the ideological barriers to these solutions. Before we focus on how we might overcome these barriers and transform some of the ideologies preventing rapid mitigation, we need to know where these dominant ideologies come from and what we are up against in trying to change them.

Often referred to as the "neoliberal" era, starting in the 1980s new ideologies permeated governance and public policy, most obviously in the United States and the United Kingdom. Ronald Reagan and Margaret Thatcher were strong proponents of "free markets" without regulation, shrinking the role of government, expanding market mechanisms as policy solutions, and attempting to curtail policies to support the environment, public health, and public education. Their administrations also worked hard to dissolve the power of unions in the workforce and dismantle forms of collective action generally. The acceptance of these policies, however, did not happen overnight. For decades, wealthy individuals and think tanks had already been spreading the ideas necessary to support neoliberal policy among the general public.

Neoliberal ideology spread through a global coordinated effort. For example, in a chapter in the book *Nine Lives of Neoliberalism*, Marie-Laure Djelic and Reza Mousavi[8] describe how a network of neoliberal think tanks, called the Atlas Network, spent decades spreading neoliberal ideology and increasing public support for neoliberal policies. Since 1981, the network has expanded to include 450 organizations in 90 different countries. Vested interests with deep pockets funded new organizations as the network spread. The primary goals were to establish influence, targeting the media and pushing forward information and narratives to shape public acceptance of neoliberal principles. By spreading these ideas, they increased public support for neoliberal principles such as a "free market," small government, reduced regulation, individualism, and a profit-oriented business mindset. The network has been highly coordinated, with organizations sharing strategies and information, and it effectively shaped ideas. As the authors state, the work of the Atlas Network promoting neoliberalism shows how

> ideas can be made to matter. Ideas do not float nor do they do things by themselves. They are championed, carried, inscribed organizationally and institutionally, fought over, appropriated, and interpreted. Only then can they come to have an impact.

This reveals that not only do ideas matter, but that powerful groups have been purposefully shaping our ideas, shifting what people think is good, desirable, acceptable, normal, natural, and possible. Fossil fuel and other vested interests have been part of this manipulation. Most importantly, these ideas did not use to be the norm but have become so due to their efforts.

In many ways the neoliberal experiment has been a failure, as it has resulted in financial crises, stagnated wages, and the first generation of Americans who are not better off than their parents. Some claim that neoliberalism is dead because it resulted in financial and social collapse (2007–2008) and staggering levels of economic inequality that continue to increase. People are doing worse, not better. Yet despite these realities, many of the core beliefs, principles, and policies remain popular. Once ideas become entrenched and normalized, it is often hard for people to let them go. Even when ideologies go against an individual's best interest, in many cases people will still cling to these beliefs because they are perceived as more certain or safe, and even as part of one's identity that connects them to other people.

In *A Theory of System Justification*, John Jost[9] explains how individuals can hold system-justifying beliefs that reinforce dominant ideologies, even when they go against their best interests. Jost argues that people have a tendency to support maintaining dominant ideas and ways of doing things due to our basic psychological motivations to (1) desire certainty about what could happen, (2) preserve a sense of safety that might be disrupted by change, and (3) maintain connections with others by sticking with the dominant ideologies associated with one's social group. These psychological motivations tend to result in people favoring keeping things the same and maintaining dominant ideologies. Groups working to maintain power can use these tendencies to their advantage, communicating in ways that make change seem uncertain, unsafe, and socially unacceptable. As Jost explains, system justification is critical to recognize because it "diminishes support for system-challenging protest activity. It is quite often an obstacle to the attainment of social justice."

System-justifying trends are even stronger in certain social contexts, for example, in a time of crisis or increased threats. In these situations, people seek certainty, safety, and social acceptance even more. Therefore, they are willing to stick with what has been the norm and resist social change, even if social changes are what is required to address the underlying drivers of the crisis. This is not inevitable, and there are many times when ideologies have shifted, but there is usually strong inherent opposition that must be recognized and counteracted with specific types of messaging.

While changing the dominant ideologies that hinder support for bold climate policies will be challenging, it is not impossible. Indeed, it is already happening slowly (more details on this below). In addition, social tipping points can also accelerate ideological transformation. We already see clear signs of these transformations taking place, and as these shifts continue they are likely to result in increased support for the transformative climate policies necessary to justly minimize warming. Embracing a more just and sustainable future means we also must embrace new ways of thinking and leave some of the old ways of thinking behind.

Shifting Dominant Ideologies

Changing our ideologies involves shifting our thinking and changing our minds in terms of what we value, what is desirable, and what is possible. In their 2023 article, social scientists Hubert Buch-Hansen and Iana Nevostra[10] highlight some of the key transformations necessary to create a more sustainable and just society. These transformations are described in terms of society needing *less* of some things and *more* of other things. For example, we need less competitiveness and greed and more solidarity and empathy. We need less waste and pollution and more regard for planetary boundaries. We need less domination and exploitation and more collaboration and equality. Because ideas matter and shape reality, changing society in these ways also requires changing the dominant ideologies reinforcing these patterns. We must shift our ideas away from some widely accepted ideas and norms and replace them with new ideas that are more conducive to a sustainable and just society. As Therborn explains, ideologies shift all the time. We must now be strategic about these shifts and guide a transition that replaces ideas that hinder creating a more sustainable and just society. Here I will highlight how we must specifically shift the dominant views about what is good and desirable in the following ways: (1) *from* individualism *to* community; (2) *from* accumulation *to* well-being; (3) *from* growth *to* sustainability; and (4) *from* extractive ownership *to* stewardship.

Individualism has become a central cultural norm in many countries due to neoliberal ideology. As Margaret Thatcher famously stated in a 1987 interview:

[T]here's no such thing as society. There are individual men and women and there are families. And no government can do anything except through people, and people must look after themselves first.

This represents one example of many where the notion of society or community is dismissed and the focus is on the individual. Individualism means that we do things for ourselves, like becoming an entrepreneur and starting our own business. It also entails not wanting government support for medical care, education, or childcare but taking care of it yourself. Individualism is often expressed in terms of freedom, freedom to do and be what you want. Individualism means we don't need community or help from other people. Sadly, it also results in widespread loneliness. Think of the lack of social interactions that come with individualism. Instead of doing something collaboratively or collectively, you do it alone. This not only means missed opportunities for social bonds but in many cases also results in more environmental harm. For example, if everyone drives their own car rather than carpooling or taking public transit together, then much more resources will be used making more cars and burning more fuel. If everyone has their own everything and we don't share, that means more material and energy use all around and more GHG emissions. It also means losing touch with the ability to share and to help one another. In an individualist world, we lose so much.

In contrast, if we replace individualism with valuing community, we see very different outcomes. In a community setting, people interact much more as they collaborate on tasks, share things, and generally interact more frequently. More people working on a task or problem is often better than going it alone. Having a group of people share things reduces material and energy use. For example, a neighborhood could share tools, lawn mowers, snowblowers, and even cars instead of everyone having their own. Embracing community, we can also carpool or take public transit with other people instead of driving alone. Talking and deliberating with others in your community also paves the way for more participatory and deeply democratic decision making. In New York City, neighborhood districts have regular meetings to bring attention to community needs and issues, and these groups even have opportunities to choose how city funds should be used in their districts. As the impacts of climate change become more severe, communities will only become more important. We cannot adapt alone. We will need each other for physical and emotional support to get through hurricanes, fires, floods, and droughts. Rebecca Solnit's book *A Paradise Built in Hell*[11] illustrates how when faced with crises and catastrophes people need each other even more, and in many cases people come together in mutual aid. Thinking in terms of a collective "we" rather than only about ourselves has many benefits and will be essential for facing the increasing impacts of climate change that we cannot avoid.

Next we need to change what we prioritize. We currently live in an economic and social system that prioritizes wealth accumulation and maximizing profits. But decisions that maximize profits also incentivize cutting costs in terms of labor, resources, and environmental protection. Making decisions based on profit has become normalized, so much so that when people or ecosystems are harmed in the name of profit maximization many people largely accept the negative impacts as "just the way things are." But prioritizing the accumulation of wealth above all other considerations has resulted in the climate and biodiversity crises as well as staggering levels of economic inequality. Average wages have remained relatively stagnant while profits and CEO salaries have soared. Those who are most wealthy have established significant influence over our political system. In summary, a society of prioritizing accumulation doesn't make us happy, undermines a sustainable future, and results in some people having tremendous power over the rest of us. Current levels of extreme wealth inequality also continue to plague society and result in social tension and political instability. Allowing our economic system to continue to prioritize profits and wealth accumulation (for the relative few) undermines well-being and is driving us toward social and environmental crises.

What if we designed society around prioritizing well-being rather than profit and wealth accumulation? This includes well-being for everyone, not just the select few. More people are talking about the idea of a "well-being economy" where decisions are based on well-being rather than financial gains. According to the Wellbeing Economy Alliance,[12] we can transform society for the better if we focus on prioritizing five aspects of a well-being economy. These include (1) *Dignity:* everyone has what is required to live in "comfort, safety, and happiness"; (2) *Fairness:* the economic system is just and equitable; (3) *Participation:* citizens participate in economic decision making; (4) *Purpose:* enhancing belonging, connection, and social support; and (5) *Nature:* living with respect for biophysical limits. Based on these principles, the scientists who authored *Earth for All*[13] developed a well-being index and stated that, beyond a global average threshold of about $15,000 per year, wealth accumulation does not increase well-being. Keep in mind that three to four billion people on the planet currently subsist on less than $4 a day. If you look closely, prioritizing wealth accumulation, or what Marjorie Kelly[14] calls "the wealth supremacy," dictates so much in our society and favors more wealth for the already wealthy. In contrast, a well-being economy would prioritize helping everyone to live with dignity, equity, and justice and would serve to reduce the drivers of the climate crisis.

In addition to abandoning wealth accumulation as the dominant social priority, we also must specifically abandon the idea that growth is good and necessary. Economic growth (in terms of GDP) as an indicator of progress causes many problems and needs to be abandoned. In wealthy countries, people are producing and consuming more per person over time. To support an always growing economy, unnecessary things are constantly being produced and marketed to consumers. A growing economy is a good thing in places where people do not have their needs met, but it is wasteful and dangerous to keep production levels high simply for the sake of growth. The growth paradigm is dangerous because it is pushing us deeper into the climate crisis. As more and more is produced, more and more GHG emissions are released. Growth in these emissions is not good. Neither is growth in extinction rates, which are also linked to GDP growth. Goods come from somewhere, and they result in pollution. Sadly, most of the world's governments still prioritize GDP growth, and many people continue to see economic growth as a sign of prosperity. Yet it is not. Countries with low GDP can have higher levels of well-being indicators than countries with high GDP (e.g., Costa Rica versus the United States). We all need to abandon the economic growth paradigm.

Instead of a growing economy, we can focus on creating a sustainable economy that doesn't push us into dangerous crises. In terms of production and consumption, that means we focus on sufficiency rather than expansion. Once needs are met, production and consumption rates can slow down to a steady state of sufficiency where everyone has enough rather than always more and more. Chapter 4 discussed specific measures that can help us to live sufficiently. Beyond specific policies we also need a sufficiency mindset. We need to be happy with having enough rather than being lured into the fantasy that having more and more will make us happier. The idea of "the good life" for many people not only includes having everything you need to live, but having more than you need: more homes, more cars, more TVs, and more stuff in general. Instead of focusing on having more to attempt to be happy, we might be happier working less and living more simply. Many people have already experimented with living simply by reducing their material belongings. Minimalism[15] is increasingly popular, and people who have adopted this mentality and lifestyle have found that it gives them more free time, reduces discontent, increases creativity, promotes connection, and provides opportunities to focus more on health and to find more purpose in life. In sum, they claim they have found *lasting happiness* through minimalism. In their view, the main argument is not that having more and more will not make you happy, but that living simply with less *will* make you happy. This mental shift is critical for low-carbon living.

Lastly, we need a mental shift in how we see nature or the environment. If you read the work of the famous scholars from the Enlightenment, you can clearly see how nature was depicted as something that should be conquered and harnessed for human use and desires. Science was also seen as a key tool to help humanity control and dominate nature. Thus a long history emerged of seeing nature as something to exploit, a commodity to be owned and extracted from to serve human interests alone. But viewing nature simply as something to own and extract from leads us right to where we are: facing environmental crises that threaten our ability to live on this planet. This outdated view of nature and the environment needs to be abandoned, as it only results in destruction. There are very real biophysical thresholds and tipping points that we are moving closer toward every year. We must reframe our view of nature.

Instead of dominating and controlling nature, using it to extract what we need and serve human interests alone, we need to adopt a more stewardship-oriented view of our environment. A 2023 United Nations review of 300 studies[16] found that Indigenous people are the best guardians of forests globally. This is because they view forests differently, not as a commodity but as a place to live and as something that must be protected for other species and future generations. As an undergraduate, I was deeply moved by the work of Aldo Leopold, who argued that we cannot continue with the mindset of nature as something to be owned and used, but must instead realize that nature is a community that we are simply one member of. As he famously stated in *A Sand County Almanac*, "a land ethic changes the role of Homo sapiens from conqueror of the land-community to plain member and citizen of it."[17] This requires a new mindset, one that Leopold argued for 75 years ago. While Leopold focused mostly on wildlife conservation, his call for a new way of thinking about nature applies even more today in terms of the climate crisis. Facing climate change and mass extinction, it seems that we have not yet understood what Leopold meant when he stated that "the conqueror role is eventually self-defeating." While many people now support sustainability broadly, we also need to think differently about nature and see ourselves as part of it. When we disrupt and push the limits of nature, we are only hurting ourselves.

These are not new ideas, and many readers may already be thinking about how these shifts represent key pillars of social transformation that have been outlined by other scholars and activists. For example, many of the things discussed above match with core principles and concepts associated with degrowth, ecosocialism, and doughnut economics. For decades people have been calling for a more sustainable and just system, and many have identified specific policies and programs that are just waiting to be implemented. While terms like degrowth might be off-putting, it is possible to focus simply on specific ideas and policies without using terms that might be misunderstood or controversial. The good thing is that even if they call it different things, many people agree on what needs to be changed in our society in terms of ideas and policies. Many also continue to work toward similar goals for positive transformation.

Building Something Better

In order to adopt the boldest climate policies with the most potential to minimize warming, the above ideological changes are key. How do we convince people to move beyond a wide acceptance of individualism, accumulation, growth, and exploiting nature? Aiming for and prioritizing these things is driving our environmental and social crises. To shift views toward seeing the benefits of community, well-being, sustainability, and stewardship, we need to be having different conversations, different media framing, and shifting values and norms. Given the widespread concern about climate change, connecting these ideological shifts to the moral imperative to minimize global warming may be easier than we think.

Despite decades of neoliberal ideology, many people are unsatisfied, lonely, and scared for the future. Many people are already seeking something more and could personally benefit from a cultural turn toward community and well-being. While neoliberal strategists harnessed the media, marketing, and scientists to make their priorities popular and common sense, similar work needs to be done to spread the word about the potential of building a more sustainable and just society that can effectively minimize global warming. The good news is that many people are already shifting the way they think, and by overcoming the silence and false assumptions about climate change, a massive shift could happen quickly.

7

Active Hope

Because we never know for sure how the future will turn out, it makes more sense to focus on what we'd like to have happen, and then to do our bit to make it more likely.

That's what Active Hope is all about.

—Joanna Macy and Chris Johnstone

While writing this book, I read a tall stack of other books focusing on different aspects of the climate crisis. A common theme emerged in the conclusions of these books. Many of them ended with a discussion of hope. Hope is critical to address the climate crisis. At this moment in history, it is hope that can guide us toward a more livable future. Not a passive form of false hope that assumes it will all work out, but an active hope that drives continued action despite what setbacks and losses might occur. Hope has always been a key ingredient for social change, and it is especially essential for pursuing change against powerful opposition.

The forces we are up against are indeed powerful, with billions of dollars at their disposal, decades of practice shaping public opinion, thousands of lawyers, hundreds of marketing teams, millions of lobbyists, and strong influence over many of our elected representatives. But we have power too. We have billions of citizens who need a sustainable future, we have hope, we have the ability to work together to demand change, and we have all the knowledge and tools we need to curb GHG emissions. The future remains uncertain, and in any uncertain situation there is always room for hope. It's time to focus on the real possibility of creating a better future. This chapter focuses on hope: the different kinds of hope, why active hope is so important, and how we can cultivate active hope as a virtue in our everyday lives. Choosing an active form of hope over doomism and despair is not only required to build a large and effective climate movement, but it is also essential for human flourishing.

False Hope

First, we must abandon all forms of false hope. False or fraudulent hope ignores reality and assumes that everything will be just fine. It is self-deception, yet it is common because it can be very comforting in troubling times. If a person has been convinced by false hope, they don't need to worry about the situation or do anything to change it. In terms of climate change, false hope also includes having faith in solutions that will be inadequate to address the crisis at hand. For example, we see false hope when people hype up "silver bullet" solutions to the climate crisis.

For example, in December 2022 the news was full of stories about an advancement in the development of fusion energy, calling it the solution to the climate crisis. Headlines included: "Fusion Breakthrough Could Be Climate, Energy Game-Changer" (Associated Press) and "Nuclear Fusion Breakthrough: Can the Quest for Clean Energy Finally Help Tackle the Climate Crisis?" (Forbes). One article even referred to fusion energy as the "vaccine" for climate change. A future powered by zero-emission fusion energy has long been pursued by scientists, yet despite optimistic goals and hopeful headlines, many obstacles remain. The breakthrough that occurred was just one of many breakthroughs that are needed to make widespread nuclear fusion energy a reality. Although they did not receive as much attention, follow-up articles focused on scientists correcting the idea that fusion could address the climate crisis. The work to make a fusion-based energy system a reality, they said, will take several decades at least. Climate change needs to be addressed now; we can't wait for a theoretical form of clean energy to eventually be developed. This is a perfect example of false hope, dangerously distracting attention away from the tools we already have and the work that must be done now to minimize warming.

Even a reliance on renewable energy as the sole solution to climate change is an example of false hope. News articles hype up renewables, citing figures of unprecedented growth in renewable production. But these articles fail to state that total energy consumption is increasing, and renewables thus far have been largely added to a growing energy system. In other words, fossil fuel use isn't going down globally, and emissions keep going up. We have to change the system that is driving this growth and undermining our best mitigation tools.

Philosopher Allen Thompson[1] states that "wishing only for an alternative energy solution, instead of making preparations for cultural change" is analogous to "turning away from reality." Thompson further explains how the focus on an alternative energy solution distracts us and prevents us from other and better possible futures. It is an appealing narrative because it provides the false assurance that there will be a "technological solution to save us all the trouble of significant behavioral and conceptual change." People keep looking for quick and easy technological solutions that allow everything else to remain the same. Yet this is a delusion, as more far-reaching social changes (like the sufficiency policies discussed in chapter 4) are required to address the climate crisis.

In addition to these forms of false hope, there is also simply passive hope. Joanna Macy and Chris Johnstone[2] explain that passive hope involves waiting for something else or someone else to bring about a desired outcome. Passive hope has no agency. For example, hoping it will not rain today is both passive and false because there are no thoughts or actions that can have a possible effect. Passive hope can also demotivate action based on the assumption that others or something else will act or be the driver of change. Stating that "I hope the climate doesn't get worse" without action, or "I hope the government adopts bold climate policy" without action, are examples of passive and false hope. As stated by Benjamin Lowe,[3] these ideas promote "a false sense of security and optimism, thus placating some and leaving them less likely to take action." False hope in all forms is dangerous because it causes passivity and inaction. Given what is at stake and our current trajectory, there is no time for passivity.

Authentic Hope Is Active Hope

In contrast to false hope, authentic hope involves acknowledging reality, the challenges ahead, and the real possibilities that lie beyond the horizon. We must honestly face reality, not deceive ourselves. With knowledge of what truly is going on, we can best assess how to move forward. As Rebecca Solnit[4] explains in her book *Hope in the Dark*, "authentic hope requires clarity—seeing the troubles in this world—and imagination, seeing what might lie beyond these situations that are perhaps not inevitable and immutable."

In terms of climate change, we need to know and acknowledge the truth. We need to know the trajectory of warming we are heading toward, what the likely impacts will be, and what tools we have to change this course.

Denying these things or trying not to think about them won't help. The key is to act on this knowledge. Hope also doesn't require any sense of optimism. One can acknowledge the reality of the situation, know the odds of success are low, and still pessimistically proceed to act. One can be pessimistic and actively working for change at the same time. This is real hope because it does not rely on false ideas or delusions of reality. Despite current trends, the future remains uncertain and unpredictable. It is still very possible to change our current climate trajectory. It is here in the realm of what is still possible that real hope resides. As explained by Rebecca Solnit[5] in *Not Too Late*, to hope is . . .

> To recognize that what is unlikely is possible, just as what is likely is not inevitable. To understand that difficult is not the same as impossible. To plan and to accept that the unexpected often disrupts plans—for the better and for the worse. To know that the powerful have their weaknesses, and we who are supposed to be weak have great power together, power to change the world, have done so before and will again.

There has been change before and there will be change again. Taking a long view of history can help to fuel hope as we acknowledge all the changes that people worked toward in the past.

Authentic hope is above all *active* and helps us to keep pursuing what we desire even if the chances of success are dim. In their book titled *Active Hope*,[6] Joanna Macy and Chris Johnstone elaborate on what active hope entails. They argue that we should not deny our feelings of grief or loss but should move through them to a place of gratitude from which we can then practice active hope. They state:

> Active Hope is a practice. Like tai chi or gardening, it is something we do rather than have. . . . First, we take in a clear view of reality; second, we identify what we hope for . . . ; and third, we take steps to move ourselves or our situation in that direction.

Active hope identifies this desired future and step by step works toward creating it. Macy and Johnstone also explain how social change often goes through three phases: (1) the idea of change is a joke; it is laughed away and dismissed as ridiculous and unrealistic; (2) the idea of change spreads and is now seen as a threat, attacked by those who oppose it; and (3) the change happens and becomes the new norm. They state that "if something is not in the picture at the moment, that doesn't mean that it won't be later on. This way of conceiving reality sees existence as an evolving story rather than as predefined." Change is inevitable, and we can actively work to direct the trajectory of this change.

Active hope is the real deal. It entails knowing the situation, knowing what you hope the future will hold, and then actively working toward that future. A tremendous amount of work is necessary to shape change and create a society with new priorities that will allow us to minimize global warming and avoid the loss and suffering associated with our current trajectory. There is so much work to be done and no time for the delusions and distraction of false hopes or the passivity of despair. If we truly want to create a more just and sustainable future, there are countless things we can be actively working toward every day.

Radical Hope: Active Hope in Dark Times

The idea of radical hope was advanced by philosopher Jonathan Lear in his 2006 book titled *Radical Hope: Ethics in the Face of Cultural Devastation*.[7] In the book, Lear examines forms of hope and courage that emerged during a time when the Crow Nation faced cultural devastation. As nomadic warriors and hunters, life for the Crow largely focused on following the buffalo and navigating intertribal warfare over territory and horses. Once forced on to their reservation, the Crow could no longer be the Crow as they were. They could no longer follow and hunt the buffalo, who were annihilated by white settlers. Neighboring tribes had also been displaced and forced into reservations. Having lost seemingly everything, what could the Crow do in the face of the collapse of their culture and way of life? Lear examines the extraordinary, radical form of hope that was demonstrated by the nation's leader, Plenty Coups, and discusses the importance of "imaginative excellence"—facing up to the challenges ahead with courage and imagining what is still possible to be gained, even when so much has been lost.

Radical hope is hope in the face of great loss. It is hope that something good will emerge, even if one does not know what that good might end up being. As Lear explains:

> What makes this hope radical is that it is directed toward a future goodness that transcends the current ability to understand what it is. Radical hope anticipates a good for which those who have the hope as yet lack the appropriate concepts with which to understand it.

When all seems lost, radical hope focuses on the possible good that could still emerge, what can be resurrected, or what new forms of human flourishing might be created.

Although there are clear differences between what the Crow Nation faced and the climate crisis, a growing number of philosophers and scholars are applying the idea of radical hope to the climate movement. Radical hope arises when the desired scenario is no longer possible. In this case, hoping for a non-warming future is pure delusion. In addition, much will be lost in terms of other species and human communities. Radical hope is authentic in that it acknowledges the truth about loss and the low odds of success but demands action because there is still something to be saved or at least the possibility that something might emerge in place of what was. Some climate activists are already drawing on radical hope to keep them active and moving forward. As one Extinction Rebellion (XR) activist explained to me, "Time is running out, but it is never too late. There will always be something left to save."

The same question that the Crow faced may soon be faced by many more communities: How can or should we live in the face of collapse? We have not yet reached total collapse, but an increasing number of communities have already faced climate-caused breakdown and there are island nations that will face certain devastation as sea level rises. In these cases, radical hope will be required to help work toward saving what can still be saved, resurrecting what can be resurrected, and creating something good, even if we don't know what that might be. Depending on a person's specific situation and outlook for the future, the hope required to keep working for positive outcomes may have to be in the form of radical hope. This is also more likely to suit those who are highly pessimistic about humanity's ability to avoid collapse.

Sociologist Carl Cassegård[8] has studied pessimistic trends within the environmental movement examining a shift from apocalyptic thinking, or focusing on *future* environmental catastrophes, toward post-apocalyptic thinking, focusing on how catastrophes are *already occurring*. Post-apocalyptic environmentalism focuses more on loss and grief, and individuals tend to turn their attention toward alternative living and the creation of new communities in transition towns or permacultures. This means individuals are already accepting the fact that catastrophe is happening and are focusing on ways to build resilience in the face of catastrophe. An example of this is the Dark Mountain Collective, which argues that while we do face collapse, future paths may open up for creating a new society. Yet these dark thoughts and the acceptance of loss do not necessarily undermine hope and action. Even when individuals are overwhelmed with grief, experiencing these emotions can actually result in "a wellspring of new forms of activism and new forms of struggles, including attempts to salvage what can still be saved."[9]

Cassegård and Thörn find that the rejection of false hope and the acceptance of catastrophe (expressed through loss and grief) can serve to motivate action: "The fact that a catastrophe appears irreparable or impossible to prevent may fuel political action . . . Out of the emotional working through of loss, new visions may emerge that can guide activism and provide blueprints for social change." Cassegård's[10] later research reveals that at times there may be ways that activists can move forward without hope and that they can find new forms of hope through action. Togetherness is a key source of hope, and lost hope can re-emerge through collective action with others. Hope still largely exists among activists, Cassegård explains, in terms of both the "big hope" of fully avoiding collapse and the "small hope" of lessening harm and seeing what can still be saved or what good can emerge out of catastrophe (radical hope). Among those who are concerned about the climate crisis and have hope, this hope may vary from big hope to small hope depending on the individual's ideas about how much catastrophe is already happening or will occur. I personally still have "big hope" that we can reduce the impacts of the climate crisis through political engagement and pressuring governments to adopt the bold climate policies we know can help minimize warming. Some people have already accepted catastrophe as inevitable and therefore must rely on radical or small hope to keep going. What matters most is that regardless of one's view on how dire things are and how deep into crisis we are going, there are forms of active hope for all of us.

Active hope is not only a key ingredient for social change but is also a key ingredient for well-being and flourishing. As philosopher Byron Williston[11] explains, sustaining well-being in the face of the climate crisis requires some amount of hope, whether big or small, and this hope can drive effective action:

> Fortunately, the hope itself can cause us to turn our energies aggressively to bringing about the outcome. This is not just because those with hope tend to generate effective routes to their goals and are good at seeing their way through or around impediments. It is also because finding a way to flourish in the teeth of the climate crisis requires working for meaning-

ful political change, acting in newly courageous ways, and looking hard
for alternative models of sustainable living.

Flourishing in the shadow of increasing climate threats requires that we
see the situation clearly and, despite the challenges, continue to act to
bring about more positive outcomes. This hope can manifest in a range
of actions from political engagement, communicating ideas and alternative ways of thinking about society, or creating strong community bonds
to weather the coming storms. The key to flourishing is to keep working
toward something better.

Active Hope as a Virtue

As described in chapter 2, virtues are qualities and characteristics that are
key to a well-lived or flourishing life. Aristotle mentioned certain virtues
to strive for and practice including moderation, justice, courage, fortitude,
modesty, truthfulness, and friendliness. Others have since come up with different lists of virtues to strive for, and some have even identified those that
are most critical to focus on to address the climate crisis. Thus we can adopt
a virtue ethics approach to the climate crisis and focus on developing these
virtues that are so critical at this moment. This includes the virtue of hope.

As mentioned in chapter 2, different climate change scholars have come
up with ideas about what virtues are essential for flourishing in the face of
global climate change. For example, Mike Hulme[12] lists key virtues that
activists should focus on, such as hope, faith, integrity, wisdom, humility,
and love. Bryon Williston names only three: hope, justice, and truth. Hope
is considered by many as a key virtue for those navigating the climate crisis.
Hope as a virtue is a practice, the practice of active hope specifically. Active
hope not only increases the likelihood of the desired outcome becoming
a reality but also serves to help individuals and groups of people come
closer to a place where they can flourish. Despair and inaction undermine
human flourishing. Active hope is the only path to the good life when so
much is at risk.

As environmental scientist Mike Hulme explains, exercising individual virtue in the face of the climate crisis requires practicing active hope.
He explains why hope is one of the most important virtues because it
"captures the human relationship with the future." The virtue of hope
expresses our relationship to the future and what is still possible. It tells us
something about the "unquenchable human desire for something better."

Hulme explains, however, that hope needs nourishing in both a personal and social context. Even when our efforts do not result in the desired outcome, we must keep practicing active hope. Virtue ethics focuses less on what we shape the world into and more on what we shape ourselves into. Knowing that you are acting virtuously can be fulfilling on its own, despite the outcome. Cultivating virtuous living is also something all of us can do, and, as Hulme believes, it might just be the way out of the climate crisis.

Byron Williston argues that one reason we have failed to address the climate crisis directly is that we have failed in terms of the three key virtues: hope, justice, and truth. Have we learned to be truly hopeful people? We do not see masses of motivated people ready to demand bold climate action. If everyone practiced the virtue of active hope, we would have many more motivated people pursuing bold climate mitigation and adaptation. Perhaps it is the lack of cultivating hope and other virtues that has resulted in the sleeping masses who fail to see that a better future is still possible and that we need action now to get there.

Virtues can be an essential guide to maintaining activism. In my own research I found that many activists were participating in XR not because they thought XR would succeed, but because they believed it was the right thing to do regardless of the outcome. They continue because they feel it is morally unacceptable to give up. The XR activists interviewed expressed that as long as there is some possibility of saving something and creating a "less bad" future, the right thing to do is act. Even while knowing the chances of their demands being met are low, they see no reason to give up. As Lowe[13] states, "a virtue ethics approach is strategic because it focuses on making us better people, regardless of the outcomes."

Philosopher Brian Treanor[14] explains the reinforcing relationship between hope and other virtues: "hopeful people are more likely to be engaged members of their communities, cultivating various public and political virtues; and being an engaged member of the community, cultivating public and political virtues, is likely to make one more hopeful." As stated by Treanor, "Today, the need for political activism is so urgent, so dire, that it fundamentally changes the kind of person we need to be in order to flourish." As many scholars have proposed, focusing on cultivating virtues and using those virtues to work toward positive outcomes offers a pathway for living the most meaningful and satisfying life we can in the face of crisis. I agree with these scholars that virtue ethics, and practicing virtues like active hope, are critical in this precarious moment in time with so much uncertainty and so much at stake.

Doomism Versus Active Hope

Doomism is a form of despair. Active hope is the opposite of despair. We cannot be doomers and practice active hope. If you have to choose one, based on the science you should reject doomism and choose hope. If you have to choose one, based on your own well-being and the prospects for social flourishing, you should reject doomism and choose hope. It is understandable to feel emotions associated with loss, grief, and even despair, but we cannot let a feeling of despair become a state of being. Despair undermines human flourishing in any context. This is why the trend of doomism is so alarming. What must so many people be thinking and feeling to result in this trend? Active hope is the antidote to doomism, and it must spread. Whether you think we are already deep into climate catastrophe or not, active and radical hope can fuel action.

We also need to keep in mind that setbacks and failures are part of social change. The lack of climate action among global governments is appalling and immoral. Some actions have been taken and they help, but they are not nearly enough to avoid massive loss and suffering. Yet change doesn't happen at a consistent pace or in only one direction. Change can involve two steps forward and one step back, or it can involve hardly any steps forward followed by a sudden cascade of bold transformation. Small events can trigger big changes. Hence the scientific agreement on the existence of both biophysical and social tipping points.

Active hope is a requirement to push society to the tipping point of bold climate action. As stated by Rebecca Solnit,[15] "we are in a climate emergency . . . [and] hope is the axe you break the door down with in an emergency." Let's get out of this emergency. Cultivating active hope is the only way forward, and I believe there is already more hope out there than we think.

Conclusion

Our Uncertain Future

The ultimate, hidden truth of the world, is that it is something we make, and could just as easily make differently.
—David Graeber

This book was inspired by the junior- and senior-level undergraduate students in my courses who I noticed had succumbed to doomism and therefore had become resigned about climate change and depressed about their future. As described in chapter 1, the science does not support doomism. There are many ways we can still lower emissions and limit global warming. Also, we are not locked in to long-term warming, as research shows temperatures will stop increasing soon after we reach net-zero emissions. Doomism might still seem attractive to some people because it justifies doing nothing. There is tremendous work to be done, and doing this work will indeed be much harder than resignation and apathy. Yet it will be worth the effort.

Even with *major* setbacks, such as national leaders being elected who want to prioritize fossil fuel extraction, there is still no reason to give up. We are not doomed, even if we face a period where policies and actions are going in the wrong direction. Perhaps such setbacks can serve to inspire bold climate policy at other levels of governance and result in other countries stepping up to illustrate what being a global climate leader looks like. Maybe it will ultimately illustrate how completely immoral it is to prioritize profits for fossil fuel interests and the wealthy while failing to protect people and robbing young people of a safer and more livable future. We cannot predict the future, but even if we go backwards for a few years, there is still no reason for despair. It remains true that there is still so much to save, and every fraction of a degree of warming avoided still matters. Setbacks are inevitable, and we must take the long view on social change. History shows us that positive social change doesn't happen at a steady pace, it is not linear, and it is often triggered by unpredictable and surprising events. And remember, in the face of setbacks, virtue ethics offers a way to maintain personal well-being and mental health. Do what is right, despite the outcome. Be true to your core values and keep working to make things better.

While bleak future climate scenarios understandably cause fear, denial, and paralysis, we must not succumb to inaction, especially when there is still so much left to save. We now have the opportunity not only to act to reduce climate-related loss and suffering, but also to work together to create something new and better. As explained by Albert Camus[1] upon reflecting on his community's response to Nazi occupation during World War II, it is through working with others for the purpose of good—namely reducing suffering and defending human dignity and justice—that humans create meaning in an otherwise absurd and indifferent world.

The future is uncertain. Yet this uncertainty is where active hope for something better lies. Thinking about a better future, talking about it, and then working toward it is key. And continue to remind yourself that uncertainty in this case is a good thing, leaving the future open to directing. As brilliantly stated by Rebecca Solnit,[2]

> We are always living in a wild future, a future sometimes inconceivably better, sometimes unimaginably worse . . . these futures that become the present were made by our actions and inactions. At best by notable individuals, but also by communities, movements, . . . people who worked together toward what seemed at first impossible, then unlikely, and then one day became the order of things.

Notes

Introduction

1. Hansen, J. E., Kharecha, P., Sato, M., Tselioudis, G., Kelly, J., Bauer, S. E., . . . & Pokela, A. 2025. Global Warming Has Accelerated: Are the United Nations and the Public Well-Informed? *Environment: Science and Policy for Sustainable Development* 67 (1): 6–44.

2. See Climate Action Tracker 2023 and the UNEP 2020 Emissions Gap Report.

3. Lee, H. et al. 2023. Synthesis Report of the IPCC Sixth Assessment Report (AR6). IPCC.

4. Bressler, R. D. 2021. The Mortality Cost of Carbon. *Nature Communications* 12 (1): 4467.

5. Wallace-Wells, D. 2018. *The Uninhabitable Earth: Life after Warming*. Time Duggan Books.

6. Hayhoe, K. 2021. *Saving Us: A Climate Scientist's Case for Hope and Healing in a Divided World*. Simon & Schuster.

Chapter 1

1. Leiserowitz, A., et al. 2023. Climate Change in the American Mind: Beliefs & Attitudes. *Yale Program on Climate Change Communication*. Fall.

2. Norgaard, K. M. 2011. *Living in Denial: Climate Change, Emotions, and Everyday Life*. MIT Press.

3. Doan, M. D. 2014. Climate Change and Complacency. *Hypatia* 29 (3): 634–50.

4. Peeters, W., Diependaele, L., and Sterckx, S. 2019. Moral Disengagement and the Motivational Gap in Climate Change. *Ethical Theory and Moral Practice* 22: 425–47.

5. Hickman, C., et al. 2021. Climate Anxiety in Children and Young People and Their Beliefs about Government Responses to Climate Change: A Global Survey. *The Lancet: Planetary Health* 5 (12): e863–e873.

6. Scranton, R. 2018. *We're Doomed. Now What?* Soho Press.

7. Lear, J. 2022. *Imagining the End: Mourning and Ethical Life*. Harvard University Press.

8. Hayhoe, K. 2021. *Saving Us: A Climate Scientist's Case for Hope and Healing in a Divided World.* Simon & Schuster.

9. Lenton, T. M., et al. 2023. *The Global Tipping Points Report 2023.* University of Exeter.

10. Osaka, S. 2023. Why Climate Doomers Are Replacing Climate Deniers. *Washington Post,* March 24.

11. Mann, M. 2022. The Best Climate Science You've Never Heard Of. https://michaelmann.net/

12. Mann, M. 2022. How a Little Discussed Revision of the Science Could Avert Doom. *Washington Post,* February 23.

13. Climate Vulnerability Monitor. 2023. https://climatevulnerabilitymonitor.org/health/

14. Bressler, R. D. 2021. The Mortality Cost of Carbon. *Nature Communications* 12 (1): L 4467.

15. Bressler, R. D., et al. 2021. Estimates of Country Level Temperature-Related Mortality Damage functions. *Scientific Reports* 11 (1): 20282.

16. Carleton, T., et al. 2022. Valuing the Global Mortality Consequences of Climate Change Accounting for Adaptation Costs and Benefits. *Quarterly Journal of Economics* 137 (4): 2037–2105.

17. Nussbaum, M. C. 2000. *Women and Human Development: The Capabilities Approach.* Cambridge University Press.

18. Lee, H., et al. 2023. Synthesis Report of the IPCC Sixth Assessment Report (AR6). IPCC.

19. See Hsiang, S., et al. 2017. Estimating Economic Damage from Climate Change in the United States. *Science* 356 (6345): 1362–69, and Calel, R., et al. 2020. Temperature Variability Implies Greater Economic Damages from Climate Change. *Nature Communications* 11 (1): 5028.

20. Nuccitelli, D. 2020. Fighting Climate Change: Cheaper Than 'Business as Usual' and Better for the Economy. *Yale Climate Connection,* November 20.

21. Köberle, A. C., et al. 2021. The Cost of Mitigation Revisited. *Nature Climate Change* 11 (12): 1035–45.

22. Treanor, B. 2010. Environmentalism and Public Virtue. *Journal of Agricultural and Environmental Ethics* 23: 9–28.

Chapter 2

1. Sparkman, G., Geiger, N. and Weber, E. U. 2022. Americans Experience a False Social Reality by Underestimating Popular Climate Policy Support by Nearly Half. *Nature Communications* 13 (1): 4779.

2. Marshall, J. et al. 2023. *Later Is Too Late.* https://potentialenergycoalition.org/global-report/

3. Ettinger, J. et al. 2023. Breaking the Climate Spiral of Silence: Lessons from a COP26 Climate Conversations Campaign. *Climatic Change* 176 (3): 22.

4. Markowitz, E. M., and Shariff, A. F. 2012. Climate Change and Moral Judgement. *Nature Climate Change* 2 (4): 243–47.

5. Jamieson, D. 2014. *Reason in a Dark Time: Why the Struggle against Climate Change Failed—And What It Means for Our Future.* Oxford University Press.

6. Gardiner, S. M. 2011. *A Perfect Moral Storm: The Ethical Tragedy of Climate Change.* Oxford University Press.

7. Cianconi, P., et al. 2023. Eco-Emotions and Psychoterratic Syndromes: Reshaping Mental Health Assessment under Climate Change. *Yale Journal of Biology and Medicine* 96 (2): 211–26.

8. See Stanley, S. K., et al. 2021. From Anger to Action: Differential Impacts of Eco-Anxiety, Eco-Depression, and Eco-Anger on Climate Action and Wellbeing. *Journal of Climate Change and Health* 1:100003, and Contreras, A., et al. 2023. When Eco-Anger (but Not Eco-Anxiety nor Eco-Sadness) Makes You Change! A Temporal Network Approach to the Emotional Experience of Climate Change. *Journal of Anxiety Disorders* 102: 102822.

9. Wray, B. 2022. *Generation Dread: Finding Purpose in an Age of Climate Anxiety.* The Experiment.

10. Harth, N. S. 2021. Affect, (Group-Based) Emotions, and Climate Change Action. *Current Opinion in Psychology* 42: 140–44.

11. Leiserowitz, A., et al. 2023. Climate Change in the American Mind: Beliefs & Attitudes. *Yale Program on Climate Change Communication.* Fall.

12. Hansen, J. E. 2009. *Storms of My Grandchildren: The Truth about the Coming Climate Catastrophe and our Last Chance to Save Humanity.* Bloomsbury.

13. Cripps, E. 2023. *Parenting on Earth: A Philosopher's Guide to Doing Right by Your Kids and Everyone Else.* MIT Press.

14. Rawls, J., 1971. *A Theory of Justice.* Belknap Press of Harvard University Press.

15. Shue, H. 2010. Policy Responses to Climate Change. In *Climate Ethics: Essential Readings*, ed. Gardiner, S. Oxford University Press.

16. Treanor, B., 2014. *Emplotting Virtue: A Narrative Approach to Environmental Virtue Ethics.* State University of New York Press.

17. Jamieson, D. 2014. *Reason in a Dark Time: Why the Struggle against Climate Change Failed—And What It Means for Our Future.* Oxford University Press.

18. Hulme, M. 2014. Climate Change and Virtue: An Apologetic. *Humanities* 3 (3): 299–312.

19. Williston, B. 2015. *The Anthropocene Project: Virtue in the Age of Climate Change.* Oxford University Press.

20. Stuart, D. 2020. Radical Hope: Truth, Virtue, and Hope for What Is Left in Extinction Rebellion. *Journal of Agricultural and Environmental Ethics* 33 (3–6): 487–504.

21. Harth, N. S. 2021. Affect, (Group-Based) Emotions, and Climate Change Action. *Current Opinion in Psychology* 42: 140–44.

22. Camus, A. 2020. *The Plague.* Penguin Classics.

23. Camus, A. 2012. *The Rebel: An Essay on Man in Revolt.* Vintage.

24. Tikkanen, R., et al. 2020. *Mental Health Conditions and Substance Use: Comparing U.S. Needs and Treatment Capacity with Those in Other High-Income Countries.* The Commonwealth Fund.

25. Patel, V., et al. 2018. Income Inequality and Depression: A Systematic Review

and Meta-Analysis of the Association and a Scoping Review of Mechanisms. *World Psychiatry* 17 (1): 76–89.

26. Bain, P. G., and Bongiorno, R. 2020. It's Not Too Late to Do the Right Thing: Moral Motivations for Climate Change Action. *Wiley Interdisciplinary Reviews: Climate Change* 11 (1): 615.

27. Bain, P.G., et al. 2016. Co-Benefits of Addressing Climate Change Can Motivate Action around the World. *Nature Climate Change* 6 (2): 154–57.

28. Klein, N. 2015. *This Changes Everything: Capitalism vs. The Climate*. Simon & Schuster.

29. Shue, H. 2021. *The Pivotal Generation: Why We Have a Moral Responsibility to Slow Climate Change Right Now*. Princeton University Press.

Chapter 3

1. Hickel, J., and Kallis, G. 2020. Is Green Growth Possible? *New Political Economy* 25 (4): 469–86.

2. Heede, R. 2014. Tracing Anthropogenic Carbon Dioxide and Methane Emissions to Fossil Fuel and Cement Producers, 1854–2010. *Climatic Change* 122 (1–2): 229–41.

3. Moran, D., et al. 2020. Quantifying the Potential for Consumer-Oriented Policy to Reduce European and Foreign Carbon Emissions. *Climate Policy* 20: S28–S38.

4. Williamson, K., et al. 2018. *Climate Change Needs Behavior Change: Making the Case for Behavioral Solutions to Reduce Global Warming*. Rare.

5. Harvey, F. 2020. Lockdowns Trigger Dramatic Fall in Global Carbon Emissions. *The Guardian*, May 19.

6. Knights, P. 2019. Inconsequential Contributions to Global Environmental Problems: A Virtue Ethics Account. *Journal of Agricultural & Environmental Ethics* 32 (4): 527–45.

7. Baatz, C., and L. Voget-Kleschin. 2019. Individuals' Contributions to Harmful Climate Change: The Fair Share Argument Restated. *Journal of Agricultural & Environmental Ethics* 32 (4): 569–90.

8. Wynes, S., and Nicholas, K. A. 2017. The Climate Mitigation Gap: Education and Government Recommendations Miss the Most Effective Individual Actions. *Environmental Research Letters* 12: 074024.

9. Perkins, S. 2017. The Best Way to Reduce Your Carbon Footprint Is One the Government Isn't Telling You About. *Science*, July 11.

10. The Energy Institute. See https://www.energyinst.org/statistical-review

11. US Energy Information Administration. See https://www.eia.gov/todayinenergy/detail.php?id=55960

12. US Energy Information Administration. See https://www.eia.gov/energyexplained/electricity/

13. See York, R. 2016. Decarbonizing the Energy Supply May Increase Energy Demand. *Sociology of Development* 2 (3): 265–72, and York, R., and Bell, S. E. 2019. Energy Transitions or Additions? Why a Transition from Fossil Fuels Requires More Than the Growth of Renewable Energy. *Energy Research & Social Science* 51: 40–43.

14. Thombs, R. 2017. The Paradoxical Relationship between Renewable Energy and Economic Growth: A Cross-National Panel Study, 1990–2013. *Journal of World-Systems Research* 23 (2): 540–64.

15. International Energy Agency. See https://www.iea.org/energy-system/energy-efficiency-and-demand/energy-efficiency

16. York, R., Adua, L., and Clark, B. 2022. The Rebound Effect and the Challenge of Moving beyond Fossil Fuels: A Review of Empirical and Theoretical Research. *Wiley Interdisciplinary Reviews-Climate Change* 13 (4): 13.

17. International Monetary Fund. 2023. See https://www.imf.org/en/Blogs/Articles/2023/08/24/fossil-fuel-subsidies-surged-to-record-7-trillion

18. International Monetary Fund. 2023. See https://www.imf.org/en/Blogs/Articles/2023/08/24/fossil-fuel-subsidies-surged-to-record-7-trillion

19. Damania, R., et al. 2023. *Detox Development: Repurposing Environmentally Harmful Subsidies.* World Bank.

20. International Institute for Sustainable Development. 2023. See https://www.iisd.org/publications/report/fanning-flames-g20-support-of-fossil-fuels

21. Milman, O. 2023. Monster Profits for Energy Giants Reveal a Self-Destructive Fossil Fuel Resurgence. *The Guardian*, February 9.

22. Laker, B. 2023. Greenwashing Unmasked: A Critical Examination of ESG Ratings and Actual Environmental Footprint. *Forbes*, August 4.

23. International Energy Agency. See https://www.iea.org/reports/net-zero-by-2050

24. Global Oil and Gas Exit List. See https://gogel.org/

25. Allen, T., and Coffin, M. 2022. Paris Maligned. Carbon Tracker Initiative. See https://carbontracker.org/reports/paris-maligned/

26. Oil Change International. 2023. Planet Wreckers Report. September 12, 2023.

27. International Energy Agency. See https://www.iea.org/reports/net-zero-by-2050

28. Bureau of Ocean Energy Management. 2023. See https://www.boem.gov/oil-gas-energy/national-program/national-ocs-oil-and-gas-leasing-program

29. ActionAid International. 2023. See https://actionaid.org/news/2023/causes-fueling-climate-crisis-are-receiving-20-times-more-financing-solutions-new

30. Harvey, F. 2023. Banks Still Investing Heavily in Fossil Fuels Despite Net Zero Pledges—Study. *The Guardian*, January 17.

31. McCulloch, N. 2023. *Ending Fossil Fuel Subsidies: The Politics of Saving the Planet.* Practical Action.

32. Semieniuk, G., et al. 2023. Potential Pension Fund Losses Should Not Deter High-Income Countries from Bold Climate Action. *Joule* 7 (7): 1383–87.

Chapter 4

1. Richardson, K., et al. 2023. Earth beyond Six of Nine Planetary Boundaries. *Science Advances* 9 (37): 2458.

2. Rachlinski, J. J. 2000. The Psychology of Global Climate Change. *University of Illinois Law Review* 299, and Baumeister, R. F., et al. 2001. Bad Is Stronger Than Good. *Review of General Psychology* 5: 323–70.

3. Chancel, L. 2022. Global Carbon Inequality over 1990–2019. *Nature Sustainability* 5 (11): 931–38.

4. Oxfam. 2023. Survival of the Richest. https://policy-practice.oxfam.org/resour ces/survival-of-the-richest-how-we-must-tax-the-super-rich-now-to-fight-inequality -621477/

5. Gossling, S., and Humpe, A. 2023. Millionaire Spending Incompatible with 1.5 degrees. *Cleaner Production Letters* 4: 1000027.

6. Oxfam International. 2023. Press release: Richest 1% bag nearly twice as much wealth as the rest of the world put together over the past two years. January 16. https:// www.oxfam.org/

7. Pew Research Center. 2020. Most Americans say there is too much economic inequality in the US, but fewer than half call it a top priority. January 9. www.pewr esearch.org

8. See https://www.inequalitymedia.org/how-wealth-inequality-spiraled-out-of -control?

9. Taffel, S. 2022. AirPods and the Earth: Digital Technologies, Planned Obsoles-cence and the Capitalocene. *Environment and Planning E: Nature and Space* 22.

10. Maitre-Ekern, E., and Dalhammar, C. 2016. Regulating Planned Obsolescence: A Review of Legal Approaches to Increase Product Durability and Reparability in Europe. *Review of European Comparative & International Environmental Law* 25 (3): 378–94.

11. Castro-Santa, J., et al. 2023. Nudging Low-Carbon Consumption through Advertising and Social Norms. *Journal of Behavioral and Experimental Economics* 102: 101956.

12. Nyfors, T., et al. 2020. Ecological Sufficiency in Climate Policy: Towards Poli-cies for Recomposing Consumption. *Futura* 3: 30–40.

13. Galbraith, J. K. 1958. *The Affluent Society*. Houghton Mifflin Harcourt.

14. Marcuse, H. 1964. *One-Dimensional Man*. Beacon Press.

15. Debord, G. 1983. *Society of the Spectacle*. Black and Red.

16. Marcuse, H. 1964. *One-Dimensional Man*. Beacon Press.

17. For example: Fitzgerald, J. B., et al. 2015. Energy Consumption and Working Hours: A Longitudinal Study of Developed and Developing Nations, 1990–2008. *Environmental Sociology* 3 (1): 213–23; Fitzgerald, J. B., et al. 2018. Working Hours and Carbon Dioxide Emissions in the United States, 2007–2013. *Social Forces* 96 (4): 1851–74; Knight, K., et al. 2013. Could Working Less Reduce Pressures on the Environment? A Cross-National Panel Analysis of OECD Countries, 1970–2007. *Global Environmental Change-Human and Policy Dimensions* 23 (4): 691–700.

18. Rosnick, D., and Weisbrot, M. 2006. *Are Shorter Working Hours Good for the Environment? A Comparison of U.S. and European Energy Consumption*. Center for Economic and Policy Research.

19. See Fitzgerald et al. 2018 above.

20. Fitzgerald, J. B. 2022. Working Time, Inequality and Carbon Emissions in the United States: A Multi-Dividend Approach to Climate Change Mitigation. *Energy Research & Social Science*. https://doi.org/10.1016/j.erss.2021.102385

21. Quinton, A. 2019. Cows and Climate Change. See www.ucdavis.edu/food/ne ws/making-cattle-more-sustainable

22. For examples, see: Eisen, M., and Brown, O. 2022. Rapid Global Phaseout of

Animal Agriculture Has the Potential to Stabilize Greenhouse Gas Levels for 30 Years and Offset 68 Percent of CO2 Emissions This Century. *PLOS Climate*; Heller, M., et al. 2020. Implications of Future US Diet Scenarios on Greenhouse Gas Emissions. CSS Report, University of Michigan, 1–24; Sun Z., et al. 2022. Dietary Change in High-Income Nations Alone Can Lead to Substantial Double Climate Dividend. *Nature Food* 3: 29–37.

23. Saving the Planet: The Market for Sustainable Meat Alternatives. 2015. UC Berkeley.

24. Vallone, S., and Lambin, E. F. 2023. Public Policies and Vested Interests Preserve the Animal Farming Status Quo at the Expense of Animal Product Analogs. *One Earth* 6 (9): 1213–26.

25. Monbiot, G. 2018. The Best Way to Save the Planet? Drop Meat and Dairy. *The Guardian*, June 8.

Chapter 5

1. Gayle, D. 2023. More Than 1,000 Climate Scientists Urge Public to Become Activists. *The Guardian*, December 4.

2. See Wright, E. O. 2010. *Envisioning Real Utopias*. Verso; and Wright, E. O. 2019. *How to Be an Anticapitalist in the Twenty-First Century*. Verso.

3. Piven, F. F. 2008. Can Power from Below Change the World? *American Sociological Review* 73 (1): 1–14.

4. Taylor, M. 2021. Environment Protest Being Criminalised around World, Say Experts. *The Guardian*, April 29.

5. Gunderson, R., and Charles, W. 2023. A Sociology of "Climatage": The Appeal and Counterproductivity of Property Destruction as a Climate Change Strategy. *Environmental Sociology* 9 (4): 398–408.

6. Foster, J. B. 2022. *Capitalism in the Anthropocene: Ecological Ruin or Ecological Revolution*. Monthly Review Press.

7. Wright, E. O. 2019. *How to Be an Anticapitalist in the Twenty-First Century*. Verso.

8. Beinhocker, E. 2019. I Am a Carbon Abolitionist. *Democracy*, June 24.

9. Otto, I., et al. 2020. Social Tipping Dynamics for Stabilizing Earth's Climate by 2050. *PNAS* 117 (5): 2354–65.

10. Cripps, E. 2023. *Parenting on Earth: A Philosopher's Guide to Doing Right by Your Kids and Everyone Else*. MIT Press.

Chapter 6

1. Sparkman, G., et al. 2022. Americans Experience a False Social Reality by Underestimating Popular Climate Policy Support by Nearly Half. *Nature Communications* 13: 4779.

2. Potential Energy Coalition. 2023. See https://potentialenergycoalition.org/global-report/

3. Sunstein, C. 2019. *How Change Happens*. MIT Press.

4. Ettinger, J., et al. 2023. Breaking the Climate Spiral of Silence: Lessons from a COP26 Climate Conversations Campaign. *Climatic Change* 176 (3): 22.

5. McRaney, D. 2022. *How Minds Change*. Penguin Press.

6. Wray, B. 2023. *Generation Dread: Finding Purpose in an Age of Climate Anxiety*. The Experiment.

7. Therborn, G. 1999. *The Ideology of Power and the Power of Ideology*. Verso.

8. Djelic, M. L., and Mousavi, R. 2020. How the Neoliberal Think Tank Went Global: The Atlas Network, 1981 to the Present. In *Nine Lives of Neoliberalism*, ed. Plehwe, D., Slobodian, Q. and Mirowski, P. Verso.

9. Jost, J. T. 2020. *A Theory of System Justification*. Harvard University Press.

10. Buch-Hansen, H., and Nesterova, I. 2023. Less and More: Conceptualising Degrowth Transformations. *Ecological Economics* 205: 107731.

11. Solnit, R. 2010. *A Paradise Built in Hell: The Extraordinary Communities That Arise in Disaster*. Penguin.

12. Wellbeing Economy Alliance. 2023. See https://weall.org/what-is-wellbeing-ec onomy

13. Dixson-Dcleve, S., et al. 2022. *Earth for All: A Survival Guide for Humanity*. New Society.

14. Kelly, M. 2023. *The Wealth Supremacy*. Barrett Koehler.

15. The Minimalists. See https://www.theminimalists.com/minimalism/

16. Carrington, D. 2021. Indigenous Peoples by Far the Best Guardians of Forests—UN Report. *The Guardian*, March 25.

17. Leopold, A. 1949. The Land Ethic. In *A Sand County Almanac and Sketches Here and There*. Oxford University Press.

Chapter 7

1. Thompson, A. 2010. Radical Hope for Living Well in a Warmer World. *Journal of Agricultural and Environmental Ethics* 23: 43–59.

2. Macy, J., and Johnstone, C. 2012. *Active Hope: How to Face the Mess We're in without Going Crazy*. New World Library.

3. Lowe, B. 2019. Ethics in the Anthropocene: Moral Responses to the Climate Crisis. *Journal of Agricultural and Environmental Ethics* 32: 479–85.

4. Solnit, R. 2016. *Hope in the Dark: Untold Histories, Wild Possibilities*. Haymarket Books.

5. Solnit, R., et al. 2023. *Not Too Late: Changing the Climate Story from Despair to Possibility*. Haymarket Books.

6. Macy, J., and Johnstone, C. 2012. *Active Hope: How to Face the Mess We're in without Going Crazy*. New World Library.

7. Lear, J. 2006. *Radical Hope, Ethics in the Face of Cultural Devastation*. Harvard University Press.

8. Cassegård, C. 2022. The Future of Environmental Movements. In *The Routledge Handbook of Environmental Movements*. Routledge.

9. Cassegård, C., and Thörn, H. 2018. Toward a Postapocalyptic Environmentalism? Responses to Loss and Visions of the Future in Climate Activism. *Environment and Planning E: Nature and Space* 1 (4): 561–78.

10. Cassegård, C. 2023. Activism without Hope? Four Varieties of Postapocalyptic Environmentalism. *Environmental Politics*. Online first.

11. Williston, B. 2012. Climate Change and Radical Hope. *Ethics and the Environment* 17 (2): 165–86.

12. Hulme, M. 2014. Climate Change and Virtue: An Apologetic. *Humanities* 3: 299–312.

13. Lowe, B. 2019. Ethics in the Anthropocene: Moral Responses to the Climate Crisis. *Journal of Agricultural and Environmental Ethics* 32: 479–85.

14. Treanor, B. 2010. Environmentalism and Public Virtue. *Journal of Agricultural and Environmental Ethics* 23 (1–2): 9–28.

15. Solnit, R., et al. 2023. *Not Too Late: Changing the Climate Story from Despair to Possibility*. Haymarket Books.

Conclusion

1. Camus, A. 2012. *The Rebel: An Essay on Man in Revolt*. Vintage.

2. Solnit, R. et al. 2023. *Not Too Late: Changing the Climate Story from Despair to Possibility*. Haymarket Books.

References

ActionAid International. 2023. https://actionaid.org/news/2023/causes-fueling-cli
mate-crisis-are-receiving-20-times-more-financing-solutions-new

Allen, T., and Coffin, M. 2022. Paris Maligned. Carbon Tracker Initiative. See
https://carbontracker.org/reports/paris-maligned/

Baatz, C., and Voget-Kleschin, L. 2019. Individuals' Contributions to Harmful
Climate Change: The Fair Share Argument Restated. *Journal of Agricultural
& Environmental Ethics* 32 (4): 569–90.

Bain, P. G., and Bongiorno, R. 2020. It's Not Too Late to Do the Right Thing:
Moral Motivations for Climate Change Action. *Wiley Interdisciplinary
Reviews: Climate Change* 11 (1): 615.

Bain, P. G., et al. 2016. Co-benefits of Addressing Climate Change Can Motivate
Action around the World. *Nature Climate Change* 6 (2): 154–57.

Baumeister, R. F., et al. 2001. Bad Is Stronger Than Good. *Review of General Psychology* 5: 323–70.

Beinhocker, E. 2019. I Am a Carbon Abolitionist. *Democracy*, June 24.

Bressler, R. D. 2021. The Mortality Cost of Carbon. *Nature Communications* 12
(1): L 4467.

Bressler, R. D., et al. 2021. Estimates of Country Level Temperature-Related Mortality Damage Functions. *Scientific Reports* 11 (1): 20282.

Buch-Hansen, H., and Nesterova, I. 2023. Less and More: Conceptualising
Degrowth Transformations. *Ecological Economics* 205: 107731.

Bureau of Ocean Energy Management. 2023. See https://www.boem.gov/oil-gas
-energy/national-program/national-ocs-oil-and-gas-leasing-program

Calel, R., et al. 2020. Temperature Variability Implies Greater Economic Damages
from Climate Change. *Nature Communications* 11 (1): 5028.

Camus, A. 2020. *The Plague*. Penguin Classics.

Camus, A. 2012. *The Rebel: An Essay on Man in Revolt*. Vintage.

Carleton, T., et al. 2022. Valuing the Global Mortality Consequences of Climate

Change Accounting for Adaptation Costs and Benefits. *Quarterly Journal of Economics* 137 (4): 2037–2105.

Carrington, D. 2021. Indigenous Peoples by Far the Best Guardians of Forests—UN Report. *The Guardian*, March 25.

Cassegård, C. 2022. The Future of Environmental Movements. In *The Routledge Handbook of Environmental Movements*. Routledge.

Cassegård C. 2023. Activism Without Hope? Four Varieties of Postapocalyptic Environmentalism. *Environmental Politics*. Online first.

Cassegård, C., and Thörn, H. 2018. Toward a Postapocalyptic Environmentalism? Responses to Loss and Visions of the Future in Climate Activism. *Environment and Planning E: Nature and Space* 1 (4): 561–78.

Castro-Santa, J., et al. 2023. Nudging Low-Carbon Consumption through Advertising and Social Norms. *Journal of Behavioral and Experimental Economics* 102: 101956.

Cianconi, P., et al. 2023. Eco-Emotions and Psychoterratic Syndromes: Reshaping Mental Health Assessment under Climate Change. *Yale Journal of Biology and Medicine* 96 (2): 211–26.

Chancel, L. 2022. Global Carbon Inequality over 1990–2019. *Nature Sustainability* 5 (11): 931–38.

Climate Action Tracker 2023, and the UNEP 2020 Emissions Gap Report.

Climate Vulnerability Monitor. 2023. https://climatevulnerabilitymonitor.org/he alth/

Contreras, A., et al. 2023. When Eco-Anger (But Not Eco-Anxiety nor Eco-Sadness) Makes You Change! A Temporal Network Approach to the Emotional Experience of Climate Change. *Journal of Anxiety Disorders* 102: 102822

Cripps, E. 2023. *Parenting on Earth: A Philosopher's Guide to Doing Right by Your Kids and Everyone Else*. MIT Press.

Damania, R., et al. 2023. *Detox Development: Repurposing Environmentally Harmful Subsidies*. World Bank.

Debord, G. 1983 [1967]. *The Society of the Spectacle*. Black and Red.

Dixson-Declève, S., et al. 2022. *Earth for All: A Survival Guide for Humanity*. New Society.

Djelic, M. L., and Mousavi, R. 2020. How the Neoliberal Think Tank Went Global: The Atlas Network, 1981 to the Present. In *Nine Lives of Neoliberalism*, ed. Plehwe, D., Slobodian, Q., and Mirowski, P. Verso.

Doan, M. D. 2014. Climate Change and Complacency. *Hypatia* 29 (3): 634–50.

Eisen, M., and Brown, O. 2022. Rapid Global Phaseout of Animal Agriculture Has the Potential to Stabilize Greenhouse Gas Levels for 30 Years and Offset 68 Percent of CO_2 Emissions This Century. *PLOS Climate*.

Ettinger, J., et al. 2023. Breaking the Climate Spiral of Silence: Lessons from a COP26 Climate Conversations Campaign. *Climatic Change* 176 (3): 22.

Fitzgerald, J. B. 2022. Working Time, Inequality and Carbon Emissions in the United States: A Multi-Dividend Approach to Climate Change Mitigation. *Energy Research & Social Science.* https://doi.org/10.1016/j.erss.2021.102385

Fitzgerald, J. B., et al. 2015. Energy Consumption and Working Hours: A Longitudinal Study of Developed and Developing Nations, 1990–2008. *Environmental Sociology* 3 (1): 213–23.

Fitzgerald, J. B., et al. 2018. Working Hours and Carbon Dioxide Emissions in the United States, 2007–2013. *Social Forces* 96 (4): 1851–74.

Foster, J. B. 2022. *Capitalism in the Anthropocene: Ecological Ruin or Ecological Revolution.* Monthly Review Press.

Galbraith, J. K. 1958. *The Affluent Society.* Houghton Mifflin Harcourt.

Gardiner, S. M. 2011. *A Perfect Moral Storm: The Ethical Tragedy of Climate Change.* Oxford University Press.

Gayle, D. 2023. More Than 1,000 Climate Scientists Urge Public to Become Activists. *The Guardian*, December 4.

Gossling, S., and Humpe, A. 2023. Millionaire Spending Incompatible with 1.5 Degrees. *Cleaner Production Letters* 4: 1000027.

Gunderson, R., and Charles, W. 2023. A Sociology of "Climatage": The Appeal and Counterproductivity of Property Destruction as a Climate Change Strategy. *Environmental Sociology* 9 (4): 398–408.

Hansen, J. E. 2009. *Storms of My Grandchildren: The Truth about the Coming Climate Catastrophe and our Last Chance to Save Humanity.* Bloomsbury.

Hansen, J. E., Kharecha, P., Sato, M., Tselioudis, G., Kelly, J., Bauer, S. E., . . . & Pokela, A. 2025. Global Warming Has Accelerated: Are the United Nations and the Public Well-Informed? *Environment: Science and Policy for Sustainable Development* 67 (1): 6–44.

Harth, N. S. 2021. Affect, (Group-Based) Emotions, and Climate Change Action. *Current Opinion in Psychology* 42: 140–44.

Harvey, F. 2020. Lockdowns Trigger Dramatic Fall in Global Carbon Emissions. *The Guardian*, May 19.

Harvey, F. 2023. Banks Still Investing Heavily in Fossil Fuels Despite Net Zero Pledges—Study. *The Guardian*, January 17.

Hayhoe, K. 2021. *Saving Us: A Climate Scientist's Case for Hope and Healing in a Divided World.* Simon & Schuster.

Heede, R. 2014. Tracing Anthropogenic Carbon Dioxide and Methane Emissions to Fossil Fuel and Cement Producers, 1854–2010. *Climatic Change* 122 (1–2): 229–41.

Heller, M., et al. 2020. Implications of Future US Diet Scenarios on Greenhouse Gas Emissions. CSS Report, University of Michigan: Ann Arbor 1–24.

Hickel, J., and Kallis, G. 2020. Is Green Growth Possible? *New Political Economy* 25 (4): 469–86.

Hickman, C., et al. 2021. Climate Anxiety in Children and Young People and Their

Beliefs about Government Responses to Climate Change: A Global Survey. *The Lancet: Planetary Health* 5 (12): e863–e873.

Hsiang, S., et al. 2017. Estimating Economic Damage from Climate Change in the United States. *Science* 356 (6345): 1362–69.

Hulme, M. 2014. Climate Change and Virtue: An Apologetic. *Humanities* 3 (3): 299–312.

International Energy Agency. 2024. Energy Efficiency. https://www.iea.org/energy-system/energy-efficiency-and-demand/energy-efficiency

International Energy Agency. 2024. Net Zero by 2050: A Roadmap for the Global Energy Sector. Published May 2021. https://www.iea.org/reports/net-zero-by-2050

International Institute for Sustainable Development. 2023. See https://www.iisd.org/publications/report/fanning-flames-g20-support-of-fossil-fuels

International Monetary Fund. 2023. https://www.imf.org/en/Blogs/Articles/2023/08/24/fossil-fuel-subsidies-surged-to-record-7-trillion

Jamieson, D. 2014. *Reason in a Dark Time: Why the Struggle against Climate Change Failed—And What It Means for Our Future.* Oxford University Press.

Jost, J. T. 2020. *A Theory of System Justification.* Harvard University Press.

Kelly, M. 2023. *The Wealth Supremacy.* Barrett Koehler.

Klein, N. 2015. *This Changes Everything: Capitalism vs. the Climate.* Simon & Schuster.

Knight, K., et al. 2013. Could Working Less Reduce Pressures on the Environment? A Cross-National Panel Analysis of OECD Countries, 1970–2007. *Global Environmental Change-Human and Policy Dimensions* 23 (4): 691–700.

Knights, P. 2019. Inconsequential Contributions to Global Environmental Problems: A Virtue Ethics Account. *Journal of Agricultural & Environmental Ethics* 32 (4): 527–45.

Köberle, A. C., et al. 2021. The Cost of Mitigation Revisited. *Nature Climate Change* 11 (12): 1035–45.

Laker, B. 2023. Greenwashing Unmasked: A Critical Examination of ESG Ratings and Actual Environmental Footprint. *Forbes*, August 4.

Lear, J. 2006. *Radical Hope, Ethics in the Face of Cultural Devastation.* Harvard University Press.

Lear, J. 2022. *Imagining the End: Mourning and Ethical Life.* Harvard University Press.

Lee, H., et al. 2023. Synthesis Report of the IPCC Sixth Assessment Report (AR6). IPCC.

Leiserowitz, A., et al. 2023. Climate Change in the American Mind: Beliefs & Attitudes. *Yale Program on Climate Change Communication.* Fall.

Lenton, T. M., et al. 2023. The Global Tipping Points Report 2023. University of Exeter.

Leopold, A. 1949. The Land Ethic. In *A Sand County Almanac and Sketches Here and There.* Oxford University Press.

Lowe, B. 2019. Ethics in the Anthropocene: Moral Responses to the Climate Crisis. *Journal of Agricultural and Environmental Ethics* 32: 479–85.

Macy, J., and Johnstone, C. 2012. *Active Hope: How to Face the Mess We're in without Going Crazy.* New World Library.

Maitre-Ekern, E., and Dalhammar, C. 2016. Regulating Planned Obsolescence: A Review of Legal Approaches to Increase Product Durability and Reparability in Europe. *Review of European Comparative & International Environmental Law* 25 (3): 378–94.

Mann, M. 2022. The Best Climate Science You've Never Heard Of. https://michaelmann.net/

Mann, M. 2022. How a Little Discussed Revision of the Science Could Avert Doom. *Washington Post*, February 23.

Marcuse, H. 1964. *One-Dimensional Man.* Beacon Press.

Markowitz, E. M., and Shariff, A. F. 2012. Climate Change and Moral Judgement. *Nature Climate Change* 2 (4): 243–47.

Marshall, J. et al. 2023. *Later Is Too Late.* https://potentialenergycoalition.org/global-report/

McCulloch, N., 2023. *Ending Fossil Fuel Subsidies: The Politics of Saving the Planet.* Practical Action.

McRaney, D. 2022. *How Minds Change.* Penguin Press.

Milman, O. 2023. Monster Profits for Energy Giants Reveal a Self-Destructive Fossil Fuel Resurgence. *The Guardian*, February 9.

Monbiot, G. 2018. The Best Way to Save the Planet? Drop Meat and Dairy. *The Guardian*, June 8.

Moran, D., et al. 2020. Quantifying the Potential for Consumer-Oriented Policy to Reduce European and Foreign Carbon Emissions. *Climate Policy* 20: S28–S38.

Norgaard, K. M. 2011. *Living in Denial: Climate Change, Emotions, and Everyday Life.* MIT Press.

Nuccitelli, D. 2020. Fighting Climate Change: Cheaper Than 'Business as Usual' and Better for the Economy. *Yale Climate Connection*, November 20.

Nussbaum, M. C. 2000. *Women and Human Development: The Capabilities Approach.* Cambridge University Press.

Nyfors, T., et al. 2020. Ecological Sufficiency in Climate Policy: Towards Policies for Recomposing Consumption. *Futura* 3: 30–40.

Oil Change International. 2023. Planet Wreckers Report. September 12.

Osaka, S. 2023. Why Climate Doomers Are Replacing Climate Deniers. *Washington Post*, March 24.

Otto, I., et al. 2020. Social Tipping Dynamics for Stabilizing Earth's Climate by 2050. *PNAS* 117 (5): 2354–65.

Oxfam. 2023. Survival of the Richest. https://policy-practice.oxfam.org/resources/survival-of-the-richest-how-we-must-tax-the-super-rich-now-to-fight-inequality-621477/

Oxfam International. 2023. Press Release: Richest 1% Bag Nearly Twice as Much Wealth as the Rest of the World Put Together over the past Two Years. January 16. https://www.oxfam.org/

Patel, V., et al. 2018. Income Inequality and Depression: A Systematic Review and Meta-Analysis of the Association and a Scoping Review of Mechanisms. *World Psychiatry* 17 (1): 76–89.

Peeters, W., Diependaele, L., and Sterckx, S. 2019. Moral Disengagement and the Motivational Gap in Climate Change. *Ethical Theory and Moral Practice* 22: 425–47.

Perkins, S. 2017. The Best Way to Reduce Your Carbon Footprint Is One the Government Isn't Telling You About. *Science.* July 11. https://www.science.org/content/article/best-way-reduce-your-carbon-footprint-one-government-isn-t-telling-you-about

Pew Research Center. 2020. Most Americans Say There Is Too Much Economic Inequality in the US, but Fewer Than Half Call It a Top Priority. January 9. www.pewresearch.org

Piven, F. F. 2008. Can Power from Below Change the World? *American Sociological Review* 73 (1): 1–14.

Potential Energy Coalition. 2023. See https://potentialenergycoalition.org/global-report/

Quinton, A. 2019. Cows and Climate Change. See www.ucdavis.edu/food/news/making-cattle-more-sustainable

Rachlinski, J.J. 2000. The Psychology of Global Climate Change. *University of Illinois Law Review* 299.

Rawls, J. 1971. *A Theory of Justice.* Belknap Press of Harvard University Press.

Richardson, K., et al. 2023. Earth Beyond Six of Nine Planetary Boundaries. *Science Advances* 9 (37): 2458.

Rosnick, D., and Weisbrot, M. 2006. Are Shorter Working Hours Good for the Environment? A Comparison of U.S. and European Energy Consumption. Center for Economic and Policy Research, December.

Scranton, R. 2018. *We're Doomed. Now What?* Soho Press.

Semieniuk, G., et al. 2023. Potential Pension Fund Losses Should Not Deter High-Income Countries from Bold Climate Action. *Joule* 7 (7): 1383–87.

Shue, H. 2010. Policy Responses to Climate Change. In *Climate Ethics: Essential Readings*, edited by Gardiner, S. Oxford University Press.

Shue, H. 2021. *The Pivotal Generation: Why We Have a Moral Responsibility to Slow Climate Change Right Now.* Princeton University Press.

Solnit, R. 2010. *A Paradise Built in Hell: The Extraordinary Communities That Arise in Disaster.* Penguin.

Solnit, R. 2016. *Hope in the Dark: Untold Histories, Wild Possibilities.* Haymarket Books.

Solnit, R., et al. 2023. *Not Too Late: Changing the Climate Story from Despair to Possibility*. Haymarket Books.

Sparkman, G., Geiger, N., and Weber, E.U. 2022. Americans Experience a False Social Reality by Underestimating Popular Climate Policy Support by Nearly Half. *Nature Communications* 13 (1): 4779.

Stanley, S. K., et al. 2021. From Anger to Action: Differential Impacts of Eco-Anxiety, Eco-Depression, and Eco-Anger on Climate Action and Wellbeing. *Journal of Climate Change and Health* 1:100003.

Stuart, D. 2020. Radical Hope: Truth, Virtue, and Hope for What Is Left in Extinction Rebellion. *Journal of Agricultural and Environmental Ethics* 33 (3–6): 487–504.

Sun, Z., et al. 2022. Dietary Change in High-Income Nations Alone Can Lead to Substantial Double Climate Dividend. *Nature Food* 3: 29–37.

Sunstein, C. 2019. *How Change Happens*. MIT Press.

Taffel, S. 2022. AirPods and the Earth: Digital Technologies, Planned Obsolescence and the Capitalocene. *Environment and Planning E: Nature and Space* 22.

Taylor, M. 2021. Environment protest being criminalised around world, say experts. *The Guardian*, April 29.

The Minimalists. See https://www.theminimalists.com/minimalism/

Therborn, G. 1999. *The Ideology of Power and the Power of Ideology*. Verso.

Thombs, R. 2017. The Paradoxical Relationship between Renewable Energy and Economic Growth: A Cross-National Panel Study, 1990–2013. *Journal of World-Systems Research* 23 (2): 540–64.

Thompson, A. 2010. Radical Hope for Living Well in a Warmer World. *Journal of Agricultural and Environmental Ethics* 23: 43–59.

Tikkanen, R., et al. 2020. Mental Health Conditions and Substance Use: Comparing U.S. Needs and Treatment Capacity with Those in Other High-Income Countries. The Commonwealth Fund.

Treanor, B. 2010. Environmentalism and Public Virtue. *Journal of Agricultural and Environmental Ethics* 23: 9–28.

Treanor, B., 2014. *Emplotting Virtue: A Narrative Approach to Environmental Virtue Ethics*. State University of New York Press.

US Energy Information Administration. See https://www.eia.gov/energyexplained /electricity/

Vallone, S., and Lambin, E. F. 2023. Public Policies and Vested Interests Preserve the Animal Farming Status Quo at the Expense of Animal Product Analogs. *One Earth* 6 (9): 1213–26.

Wallace-Wells, D. 2018. *The Uninhabitable Earth: Life after Warming*. Time Duggan Books.

Wellbeing Economy Alliance. 2023. See https://weall.org/what-is-wellbeing-eco nomy

Williamson, K., et al. 2018. *Climate Change Needs Behavior Change: Making the Case for Behavioral Solutions to Reduce Global Warming.* Rare.

Williston, B. 2012. Climate Change and Radical Hope. *Ethics and the Environment* 17 (2): 165–86.

Williston, B. 2015. *The Anthropocene Project: Virtue in the Age of Climate Change.* Oxford University Press.

Wray, B. 2022. *Generation Dread: Finding Purpose in an Age of Climate Anxiety.* The Experiment.

Wright, Erik O. 2010. *Envisioning Real Utopias.* Verso.

Wright, Erik O. 2019. *How to Be an Anticapitalist in the Twenty-First Century.* Verso.

Wynes, S., and Nicholas, K. A. 2017. The Climate Mitigation Gap: Education and Government Recommendations Miss the Most Effective Individual Actions. *Environmental Research Letters* 12: 074024.

York, R. 2016. Decarbonizing the Energy Supply May Increase Energy Demand. *Sociology of Development* 2 (3): 265–72.

York, R., Adua, L., and Clark, B. 2022. The Rebound Effect and the Challenge of Moving beyond Fossil Fuels: A Review of Empirical and Theoretical Research. *Wiley Interdisciplinary Reviews-Climate Change* 13 (4): 13.

York, R., and Bell, S. E. 2019. Energy Transitions or Additions? Why a Transition from Fossil Fuels Requires More than the Growth of Renewable Energy. *Energy Research & Social Science* 51: 40–43.

Index